Astrophysics and Space Science Library

EDITORIAL BOARD

Chairman

W.B. BURTON, *National Radio Astronomy Observatory, Charlottesville, VA, USA*
bburton@nrao.edu
University of Leiden, Leiden, The Netherlands
burton@strw.leidenuniv.nl

F. BERTOLA, *University of Padua, Padua, Italy*
J.P. CASSINELLI, *University of Wisconsin, Madison, USA*
C.J. CESARSKY, *Commission for Atomic Energy, Saclay, France*
P. EHRENFREUND, *University of Leiden, Leiden, The Netherlands*
O. ENGVOLD, *University of Oslo, Oslo, Norway*
A. HECK, *Strasbourg Astronomical Observatory, Strasbourg, France*
E.P.J. VAN DEN HEUVEL, *University of Amsterdam, Amsterdam, The Netherlands*
V.M. KASPI, *McGill University, Montreal, Canada*
J.M.E. KUIJPERS, *University of Nijmegen, Nijmegen, The Netherlands*
H. VAN DER LAAN, *University of Utrecht, Utrecht, The Netherlands*
P.G. MURDIN, *Institute of Astronomy, Cambridge, UK*
F. PACINI, *Istituto Astronomia Arcetri, Firenze, Italy*
V. RADHAKRISHNAN, *Raman Research Institute, Bangalore, India*
B.V. SOMOV, *Astronomical Institute, Moscow State University, Moscow, Russia*
R.A. SUNYAEV, *Space Research Institute, Moscow, Russia*

For further volumes:
www.springer.com/series/5664

A Brief History of Radio Astronomy in the USSR

S.Y. Braude · B.A. Dubinskii ·
N.L. Kaidanovskii · N.S. Kardashev ·
M.M. Kobrin · A.D. Kuzmin · A.P. Molchanov ·
Y.N. Pariiskii · O.N. Rzhiga ·
A.E. Salomonovich · V.A. Samanian ·
I.S. Shklovskii · R.L. Sorochenko · V.S. Troitskii
Editors

A Brief History of Radio Astronomy in the USSR

A Collection of Scientific Essays

Editor of the English Translation: Kenneth I. Kellermann
Translated by Denise C. Gabuzda

 Springer

Editors
S.Y. Braude
Institute of Radio Astronomy
National Academy of Sciences
Kharkov, Ukraine

B.A. Dubinskii
Institute of Radio Engineering and Electronics
Russian Academy of Sciences
Moscow, Russia

N.L. Kaidanovskii
Central Astronomical Observatory
Russian Academy of Sciences
St. Petersburg, Russia

N.S. Kardashev
Astro Space Center
Lebedev Physical Institute
Moscow, Russia

M.M. Kobrin
Gorkii Physical-Technical Research Institute
Radio Physical Research Institute
USSR Academy of Sciences
Gorkii, Russia

A.D. Kuzmin
Lebedev Physical Institute
Russian Academy of Sciences
Moscow, Russia

A.P. Molchanov
Central Astronomical Observatory
Russian Academy of Sciences
St. Petersburg, Russia

Y.N. Pariiskii
Central Astronomical Observatory
Russian Academy of Sciences
St. Petersburg, Russia

O.N. Rzhiga
Institute of Radio Engineering and Electronics
Russian Academy of Sciences
Moscow, Russia

A.E. Salomonovich
Lebedev Physical Institute
Russian Academy of Sciences
Moscow, Russia

V.A. Samanian
Byurakan Astrophysical Observatory
Byurakan, Aragatzotn province, Armenia

I.S. Shklovskii
Space Research Institute
Russian Academy of Sciences
Moscow, Russia

R.L. Sorochenko
Lebedev Physical Institute
Russian Academy of Sciences
Moscow, Russia

V.S. Troitskii
Radio Physical Research Institute
Nizhnii Novgorod, Russia

K.I. Kellermann
National Radio Astronomy Observatory
Charlottesville, VA, USA

ISSN 0067-0057   Astrophysics and Space Science Library
ISBN 978-94-007-2833-2   e-ISBN 978-94-007-2834-9
DOI 10.1007/978-94-007-2834-9
Springer Dordrecht Heidelberg London New York

Library of Congress Control Number: 2012933119

© Springer Science+Business Media B.V. 2012
No part of this work may be reproduced, stored in a retrieval system, or transmitted in any form or by any means, electronic, mechanical, photocopying, microfilming, recording or otherwise, without written permission from the Publisher, with the exception of any material supplied specifically for the purpose of being entered and executed on a computer system, for exclusive use by the purchaser of the work.

Cover illustration: DKR-1000 cross radio telescope of Lebedev Physical Institute (FIAN): East–West arm

Printed on acid-free paper

Springer is part of Springer Science+Business Media (www.springer.com)

Preface to English Edition

As was the case in Western countries, experimental radio astronomy in the Soviet Union largely grew out of wartime radar research programs. However, unlike in Europe, Australia, and the United States where post-war research was implemented primarily in universities and in civilian research laboratories, in the USSR any research with potential military application, such as radio and radar astronomy remained largely within military-oriented and tightly controlled laboratories. As such, publication in the open literature was restricted, and when published, important experimental details were usually omitted, so the results were often suspect or ignored by the Western scientific community. Thus, although starting in 1958, many of the most important Soviet journals were translated into English, for the most part Soviet observational radio astronomy had little impact outside of the Soviet Union. By contrast, the theoretical work of Soviet scientists such as Iosif Shklovskii, Solomon Pikel'ner, Vitaly Ginzburg, Yakov Zel'dovich in Moscow and Victor Ambartsumian in Armenia and later their students, including Nikolai Kardashev, Igor Novikov, and Vyacheslav Slysh, at the Sternberg Astronomical Institute, Yuri Pariiskii in Lenningrad, and Rashid Sunyaev at the Institute of Applied Mathematics was widely recognised and considerably influenced both theoretical thinking as well as motivating new observational radio astronomy programs in the United States, Australia, and Europe. Indeed, during the 1960s, Shklovskii's book, "Cosmic Radio Waves"[1] was widely used throughout the world by students of radio astronomy.

Radio astronomy in the USSR began with the work of Ginzburg[2] and Shklovskii[3] who independently derived the high temperature of the solar corona in 1946. Interestingly, although they were both theoreticians throughout their careers, both Ginzburg and Shklovskii travelled to Brazil as part of an early Soviet expedition to study the sun during the total eclipse of May 20, 1947. Although beleaguered by the death of expedition head, N.D. Papaleksi, just a few months prior to the eclipse and

---

[1] Shklovskii, I., *Cosmic Radio Waves,* Harvard University Press, Cambridge Massachusetts (1960).
[2] Ginzburg, V.L., 1946, Dokl. Akad. Nauk SSSR, 52, 487.
[3] Shklovskii, I.S., 1946, Astron. Zhur. 23, 333.

delayed an unusually late thawing of the winter ice their Latvian port, S. E. Khaikin and B. M. Chikhachev[4] succeeded in showing that, as predicted by Shklovskii and Ginzburg, the radio emission from the sun came from the much larger corona and not from the eclipsed solar disk (Sect. 1.1).

The early observational radio astronomy programs in the USSR were carried out primarily by people trained in physics or engineering. Many different Soviet laboratories were engaged in radio astronomy, and there was often significant competition among laboratories both for recognition and for resources. With a few exceptions, there was relatively little interaction between these radiophysicists and the more traditional Soviet astronomical community. Observational radio astronomy in the USSR was highly organised under the Scientific Council for Radio Astronomy of the USSR Academy of Sciences, for many years under the leadership of Academy Vice-President Vladimir Kotel'nikov. The main programs were centred at Lebedev Physical Institute (FIAN) (Chap. 1) with field stations in Crimea and later in Pushchino near Moscow, at the Main Astronomical Observatory in Pulkovo[5] (GAO) near Leningrad (Chap. 4), and at the Radio-Physical Research Institute (NIRFI) in Gorkii (Chap. 2) as well as at the Ukrainian Institute of Radio Physics and Electronics in Kharkov (Chap. 8). Skilled scientists, such as Vsevolod Troitskii in Gorkii, Viktor Vitkevich at FIAN, Semen Khaikin in Pulkovo, and Semen Braude in Kharkov often working with less than state-of-the art instrumentation made a number of important investigations. Somewhat later their students including Yuri Pariiskii in Pulkovo, Genadii Sholomitskii at Sternberg Astronomical Institute along with Arkadii Kuz'min, Roman Sorochenko and Leonid Matveenko at FIAN became the next generation of leaders. However, partly due to the different research cultures, as well as to the poor communication across the "Iron Curtain," to this day, their work has had little impact outside of the USSR. The translation of *A Brief History of Radio Astronomy in the USSR*, for the first time makes available in the English language, descriptions of the antennas and instrumentation used in the USSR, the astronomical discoveries, as well as interesting personal backgrounds on many of the early key players in Soviet radio astronomy.

For example we can read in Sects. 1.3 and 4.4 of the discovery of radio recombination lines by two independent laboratories at FIAN[6] and in Pulkovo[7] in 1964 and perhaps as early as 1963, although this important discovery is usually credited to Hoglund and Mezger[8] who reported the detection of the hydrogen 109α line in 1965 from a number of HII regions using the newly completed 140-ft radio telescope in Green Bank, WV. As this was the first important result from the 140-ft radio telescope, which along with a series of follow-up observations, rescued the

---

[4]Khaikin, S.E. and Chikhachev, B.M., 1947, Dokl. Akad. Nauk SSSR, 58, 1923.

[5]Pariiskii, Y.N., 2007, Astron. Nach., 328, 405.

[6]Sorochenko, R.L. and Borodzich, E.V., 1965, Dokl. Akad. Nauk SSSR, 163, 603; 1966, Soviet Phys. Dokl., 10, 588.

[7]Dravskikh, A.F., Dravskikh, Z.V., Kolbasov, V.A., Misez'hnikov, G.S., Nikulin, D.E., and Shteinshleiger, V.B., 1965, Dokl. Akad. Nauk 163, 332; 1966, Soviet Phys.-Dokl., 10, 627.

[8]Höglund, B. and Mezger, P.G., 1965, Science 150, 339.

reputation of the 140-ft antenna as well as the Green Bank Observatory following a lengthy expensive construction period, it was perhaps convenient to ignore the earlier Soviet result.

The possibility of observing large n transitions in atomic hydrogen was discussed as early as 1945 by Henk van de Hulst[9] in his now famous paper which also discusses the 21 cm hyperfine structure line from atomic hydrogen, almost as an afterthought to the more extensive discussion of free-free and recombination line emission. However, van de Hulst erroneously concluded that due to Stark broadening, radio recombination lines would not be observable. Thus, although suitable equipment existed in many laboratories around the world, it wasn't until 1958 when Nikolai Kardashev[10] published an independent analysis showing that the effects of Stark broadening may have been previously over estimated, were there any serious attempts to detect radio recombination lines. The successful independent detections of the H90α line at 8872.5 MHz by Roman Sorochenko and Eduard Borodzich using the FIAN 22-m radio telescope in Pushchino and the 104α line at 5763 MHz by Alexander Dravskikh et al. at the Main Astronomical Observatory in Pulkovo were first reported at the XII IAU General Assembly in Hamburg, Germany in August 1964. Due to travel restrictions, the papers were presented by Yuri Pariiskii and Viktor Vitkevich, respectively. However owing to a combination of language difficulties, the poor quality of the visual material, the very restricted information about the instrumentation used that was permitted by the Soviet authorities, and the social isolation of the Soviet participants from the Western radio astronomy attendees resulting from their carefully monitored activities, the Pulkovo and Pushchino discoveries were widely discounted outside the USSR. However, in mid 1980's, in recognition of this important discovery, a team, including Sorochenko, Kardashev, Borodzich, and Alexander and Zoya Dravskikh received the USSR State Prize, one of the highest marks of recognition in science in Former Soviet Union.

As discussed in Sect. 3.2, a similar situation occurred following the 1965 discovery of radio source variability by Genadii Sholomitskii, who was then a young Moscow University graduate student of Shklovskii. At the suggestion of Shklovskii, Sholomitskii used the Crimean deep space tracking antenna system near Evpatoria to discover radio variability at 30 cm wavelength with a period of about 100 days in the well known peculiar quasar CTA 102. Sholomitskii's discovery was announced in an Astronomical Telegram[11] and in a short paper in the Astronomiicheskii Zhurnal[12] which generated considerable attention in the West, but for several reasons this unexpected result was generally discounted by Western radio astronomers. First, apparently for security reasons, no experimental details were given in the published papers to substantiate the claimed results, although a picture of an unfamiliar antenna system was shown in the main Soviet daily newspaper, Pravda, which did

---

[9] Van der Hulst, H., 1945, Nederlandsch. Tijdschr. V. Natuurkunde 11, 201, see also Sullivan.

[10] Kardashev, N., 1959, Astron. Zh., 36, 838; Soviet Astron.-AJ, 3, 813.

[11] Sholomitskii, G.B., 1965, IAU Information Bulletin on Variable Stars, 83, 1.

[12] Sholomitsky, G.G., 1965, Astron. Zh. 42, 673; Soviet Astron.-AJ 9, 516.

arouse considerable interest within Western intelligence circles. Secondly, observations at several Western observatories did not show any evidence for radio variability in CTA 102 or any other radio source. Finally, and perhaps most important, it was understood by everyone, including the members of Shklovskii's group, that such rapid variability was "theoretically impossible" since light travel time arguments meant that the source would need to be so small, that any radio emission would be self absorbed.

Indeed, the theoretical objections appeared so compelling, that at a press conference at the Sternberg Institute, Kardashev half jokingly suggested that perhaps the radio emission from CTA 102 might be a transmission from an extraterrestrial intelligence. This was reported on the front page of Pravda, and was picked up by newspapers around the globe, further detracting from the credibility of the claimed variability. It was not until a few years later, when Bill Dent[13] reported observing radio variability at the University of Michigan Radio Observatory, that the phenomena of radio variability was accepted. We now know that the radio emission from CTA 102 does vary at 30 cm on the time scales reported by Sholomitskii, as do many other quasars, and that this phenomena is now understood to occur as a combination of relativistic beaming and interstellar scattering.

One Soviet observational program which was widely recognised, was the series of experiments to establish an accurate flux density scale for discrete radio sources. Although the measurement of the relative strength of discrete radio sources is straight forward, one of the outstanding challenges in experimental radio astronomy is the absolute calibration of the discrete source flux density scale. As described in Sect. 2.2.2, V. Vsevolod Troitskii and his colleagues in Gorkii carried out a series of elegant experiments using an "artificial moon" as a black body standard reference source. The results of this work on absolute calibrations were subsequently used throughout the world to calibrate relative measurements made with other facilities.

Chapter 9 discusses the planetary radar program led by Academician Vladimir Kotel'nikov which was closely coupled to the Soviet space program. Unlike the passive radio astronomy program, perhaps heightened by the existing cold-war competition, there was an intense rivalry between the Russian and American attempts to be the first to detect radar reflections from the planet Venus. The prize was not only scientific priority, but the accurate determination of the Astronomical Unit important for planned missions to Venus and Mars by both the USSR and the US. But the account given in Chap. 9 makes only passing reference to the earlier work at the Goldstone and Millstone Hill facilities in the US. Indeed, the initial Russian announcement of the value of the AU based on their 1961 measurements was remarkably close to the value that had been previously announced by Goldstone which was later recognised to be based on a spurious detection from Venus.

Section 1.3 discusses the 1962 suggestion by Leonid Matveenko on building an independent-oscillator-tape-recording interferometer which could allow the unlimited extension of interferometer baselines to gain extraordinary high angular resolution. No attempt was made to implement this new technique, probably due to the

---

[13] Dent, W., 1965, Science 148, 1458.

lack of suitably instrumentation in Russia, and the combination of Soviet bureaucracy and secrecy delayed publication of these ideas until 1965.[14] However, by this time programs were already underway in the US and Canada to implement these techniques for Very Long Baseline Interferometry. Apparently, there was some discussion with A.C.B. Lovell and Henry Palmer to develop a radio interferometer between Russia and Jodrell Bank, but nothing ever materialised from these discussions. Much later, Matveenko and others collaborated with US radio astronomers to implement independent-oscillator tape-recording interferometry between Crimea and the United States.[15] For over three decades, Russian radio astronomers, led by Academician Kardashev, have been preparing a satellite known as "RadioAstron" to go into very high orbit to enable very long baseline interferometer observations in conjunction with ground-based radio telescopes in many countries to increase the resolution over purely ground based observations by more than an order of magnitude.[16]

Outside of Russia, the most influential Soviet radio astronomy observations were based on the low frequency arrays developed by Simon Braude and later by Leonid Litvinenko and Alexander Konovalenko near the Ukrainian city of Kharkov (Chap. 8). For many years this was the most powerful facility in the world working at decameter wavelengths. Braude and his colleagues carried out a number of interesting programs on radio source spectra and high n recombination lines, but their work was plagued by the absorption and distortions which took place in the ionosphere. Only recently, with the development of sophisticated digital technology and high speed computing, are these problems being successfully attacked by the new generation of low frequency radio telescopes in Europe, the US, and in Australia which are building on the pioneering work begun at Kharkov.

It is clear from many of the accounts reported in this book that many Soviet radio astronomers appeared as unfamiliar with Western radio astronomy programs as were Western scientists about the Soviet work. Due to hard currency restrictions in effect at the time, Western journals and books were routinely copied in the USSR and were widely distributed throughout the country. However, this meant very long delays between the time of publication and when the journal became available to individual scientists. While the academic astronomers, for example at the Sternberg Institute, had a good reading knowledge of English, the observational radio astronomers, who were primarily educated as engineers, were less comfortable with English, and may not have carefully followed the foreign literature.

Perhaps because of their lack of contact with Western radio astronomers, Soviet scientists were slow to pick up the growing trend in the 1960s and 1970s to build large arrays of modest sized dish antenna such as the Westerbork Array, the VLA, and the Australia Telescope Compact Array. Instead, concerned about phase stability problems inherent in multi-element interferometer arrays, and not having

---

[14] Matveenko, L.I., Kardashev, N.S. and Sholomskii, G.V., 1965, Radiophysica, 8, 651; Soviet Radiophysics, 8, 461.

[15] Matveenko, L.I., 2007, Astron. Nach. 328, 411.

[16] Kardashev, N.S., 1999, Experimental Astronomy, 7, 329.

access to the computing facilities needed to analyse multi-element interferometer data, the Russians concentrated on large one dimensional filled aperture standing arrays or phased arrays, such as the Pulkovo antenna designed by Khaikin, and later the RATAN-600 antenna, (Chap. 4) neither of which made a big impact to radio astronomy outside the Soviet Union. The 22-m steerable radio telescope located in Pushchino, near Moscow, and its twin even more accurate, version built in Crimea on the shores of the Black Sea were, however, the first radio telescopes of their size to operate at millimetre wavelengths (Sect. 1.3). Both the Pushchino and Crimean antennas were used for some of the earliest radio observations at millimetre wavelengths, especially of the planets, but their impact was limited by the poor sensitivity of the receivers available to the Soviet radio astronomers. The Crimean antenna was used in 1969 for the first VLBI observations between the USSR and the US,[17] and later, it was used together with the European VLBI Network.

As already mentioned, Soviet theoretical work had a much greater impact in the West than the observational programs. Probably the most productive theoretical programs were those led by Shklovskii at the Moscow State University Sternberg Astronomical institute (GAISH) and later at Space Research Institute (IKI) and Ginzburg at the Lebedev Physical Institute (FIAN). Ginzburg's contributions to basic synchrotron radiation theory[18] had a big impact in the West, including his two extensive articles in the Annual Reviews of Astronomy and Astrophysics[19,20] which form the basis for our current interpretation of the synchrotron radiation from radio galaxies and quasars. Similarly the pioneering work on the early universe and large scale cosmic structures by Zel'dovich, and later his young collaborators Igor Novikov and Rashid Sunyaev was well known throughout the world and continues to this day to influence current thinking.

Shklovskii and his close group of students, who were first part of GAISH and later were located at the Space Research Institute (IKI), were arguably the world's outstanding theoretical group in radio astronomy (see Chap. 3). They not only provided innovative interpretations of the plethora of new observational discoveries being made in Europe, Australia, and the United States, but perhaps more important they predicted a number of new phenomena that could be observationally tested. Unlike their observational counterparts in the USSR, the GAISH/IKI group was well informed about what was happening outside the USSR. They were a valuable part of the international radio astronomy community and they traded letters and where possible personal visits. Shklovskii, along with his students, Slysh and Kardashev, applied the synchrotron theory to interpret the radio emission from supernovae[21] and

---

[17] Broderick, J.J. et al. 1970, Astron. Zhur. 47, 748; Soviet Astron.-AJ, 14, 627.

[18] Ginzburg, V.L., 1951, Dokl. Akad. Nauk, 76, 377; in Classics in Radio Astronomy, ed. W.T. Sullivan III (Reidel), p. 93.

[19] Ginzburg, V.L. and Syrovatskii, S.I., 1965, ARAA, 3, 297.

[20] Ginzburg, V.L. and Syrovatskii, S.I., 1969, ARAA, 7, 1969.

[21] Shklovskii, I.S., 1960, Astron. Zhur. 37, 256; Soviet Astron.-AJ, 4, 243.

radio galaxies[22,23] which motivated many new observational programs. Particularly important were Shklovskii's prediction of the M87 polarisation,[24] his prediction of the 2% per year flux density decay of the Cas A supernova remnant,[12] his calculation of the HI,[25] Deuterium,[26] OH and CH[27] line frequencies, Slysh's interpretation of the peaked spectrum radio sources CTA 21 and CTA 102 as due to synchrotron self absorption,[28] Kardashev's explanation of radio source spectra as the result of synchrotron radiation cooling,[29] and as mentioned above, his classical paper on radio recombination lines.[6] Interestingly, Shklovskii's group appeared to have little or no access or need of computers, and their analysis was characteristically reduced to simple problems that could be worked analytically. Although Shkovskii's work was widely recognised in the West, unlike his counterparts Ginzburg, Zel'dovich, and Ambartsumian, Shklovskii was never elected as a full Member of the Soviet Academy of Sciences. An account of Sklovskii's life is told in his entertaining autobiographical book, *Five Billion Vodka Bottles to the Moon: Tales of a Soviet Scientist*.[30]

Section 3.4 reports on the prescient 1964 paper by Igor Novikov and Andrei Doroshkevich[31] which predicted the existence of a cosmic microwave background (CMB) that could have been experimentally tested. Novikov and Doroshkevich were sufficiently familiar with the Western literature to realise that a relevant experiment in the United States had been previously reported by E.A. Ohm in the Bell System Technical Journal.[32] However, it remained for Penzias and Wilson,[33] who were unaware of the Novikov and Doroshkevich paper or even the Ohm paper from their own laboratory, to discover the CMB the following year and relate it to the big-bang early universe. As reported in Sect. 4.4, as early as 1959, T. Shmaonov, working at the Pulkovo Observatory, may have marginally detected the CMB as part of his PhD research using a horn antenna at 3.2 cm. However, no one in the USSR or elsewhere related this to the CMB until after learning of the Penzias and Wilson paper.

For many years Russian scientists were leading the Search for Extraterrestrial Intelligence (SETI). Kardashev's classic 1963 paper on so-called Type I, Type II,

---

[22] Shklovskii, I.S., 1962, Astron. Zh., 39, 591; 1963; Soviet Astron.-AJ, 6, 465.

[23] Shklovskii, I.S., 1963, Astron. Zhur., 40, 972; 1964, Soviet Astron.-AJ, 7, 748.

[24] Shklovskii, I.S., 1955, Astron Zhur., 32, 215.

[25] Shklovskii, I.S., 1949, Astron. Zh. 26, 10.

[26] Shklovskii, I.S., 1952, Astron. Zh. 29, 144.

[27] Shklovskii, I.S., 1949, Dokl. Akad. Nauk, 92, 25.

[28] Slysh, V., 1963, Nature 199, 682.

[29] Kardashev, N.S., 1962, Astr. Zhur. 39, 393; 1962, Soviet Astron.-AJ 6, 317.

[30] Shklovskii, I.S., 1991, Five Billion Vodka Bottles to the Moon: Tales of a Soviet Scientist (W.W. Norton & Company).

[31] Doroshkevich, A.G. and Novikov, I.D., 1964, Dokl. Akad. Nauk, 154, 809.

[32] Ohm, E.A., 1961, Bell System Technical Journal, 40, 1065.

[33] Penzias, A.A. and Wilson, R.W., 1965, ApJ 142, 419.

and Type II civilisations[34] continues to define observational SETI programs, while Shkovskii's book[35] after translation to English with Carl Sagan as a co-author became a standard reference on SETI.[36] Section 3.3 describes how the interest in pursuing an observational SETI program, led in part to the establishment of the RATAN-600 radio telescope which, since 1976, has continued to be the most powerful radio astronomy facility in Russia.

In spite of the tensions between the USSR and Western countries, scientific exchanges between the two countries began in the early 1960s. In 1961, a group of Soviet and American astronomers met in Green Bank, WV to discuss various topics in radio astronomy. In 1964, Arkadii Kuz'min from FIAN spent a year at Caltech working with Barry Clark to make interferometric studies of the planet Venus[37] which provided important guidance to the planning of both U.S. and Soviet missions to Venus. Later George Swenson (University of Illinois) and Ron Bracewell (Stanford) participated in an exchange visit to the Soviet Union. Long term visitors to the USSR included Malcolm Longair from the Cavendish Laboratory who worked closely with Zeldovich and his group, and later, Denise Gabuzda who very ably translated this volume and who worked at FIAN from 1994 to 1998, and who continues to work with both Russian graduate and undergraduate students. Starting in 1969, Russian and US radio astronomers began a collaboration in Very Long Baseline Interferometry which later included observations between European radio telescopes and the growing network of Russian radio telescopes.[14,38]

Following the difficult economic times after fall of the Soviet Union in 1991, radio astronomers from the Former Soviet Union have become more integrated with their Western counterparts with greatly improved communication and increasing ease of movement between Russian and Western scientists and students.

Unfortunately, *A Brief History of Radio Astronomy in the USSR* does not give citations to many of the original works which would have greatly added to the value of the book. Therefore, some key references are given in this section for readers that wish to learn more about early radio astronomy in the USSR. More detailed reports of early Soviet era observational and theoretical radio astronomy can be found in W.T. Sullivan's books, *The Early years of Radio Astronomy*[39] and *Cosmic Noise*.[40]

I am indebted to Denise Gabuzda, Leonid Gurvits, Yuri Kovalev, Malcolm Longair, and Leonid Matveenko, and Woody Sullivan for clarifying many points relat-

---

[34] Kardashev, N., 1963, Astr. Zh. 41, 2; 1964, Soviet Astronomy-AJ, 8, 217.

[35] Shklovskii, I.S., *The Universe, Life, and Intelligence*, USSR Academy of Science, Moscow (1961).

[36] Shklovskii, I.S. and Sagan, C., *Intelligent Life in the Universe*, Holden-Day, San Francisco (1966).

[37] Clark, B. and Kuz'min, A.D., 1964, ApJ, 142, 23.

[38] Kellermann, K.I., 1992, in Astrophysics on the Threshold of the 21st Century, ed. N.S. Kardashev (Gordon and Breach), pp. 37–51.

[39] Sullivan III, W.T. 1984, The Early years of Radio Astronomy, Cambridge University Press, pp. 268–302.

[40] Sullivan III, W.T. 2009, Cosmic Noise, Cambridge University Press, pp. 214–221, 380–384.

ing to the history of radio astronomy in the Former Soviet Union. Leonid Gurvits, Yuri Ilyasov, Leonid Lytvynenko, and Leonid Matveenko provided high resolution copies of many of the illustrations which are reproduced in this English edition, in place of the poor quality originals Russian volume. In some cases, for clarity, we have substituted, similar, although not identical, illustrations when there was no adequate version of the original.

Charlottesville, VA, USA   K.I. Kellermann

Preface

The first results of observations of radio waves arriving at the Earth from the cosmos were published in 1932. These observations, which were carried out at the end of 1931 by the American engineer Karl Jansky, who was studying radio interference at 14.6 m, marked the birth of radio astronomy.

The 50th birthday of radio astronomy led to a growth in interest in both radio astronomy results and the history of the birth and development of this new scientific field.

Radio astronomy began to be intensely developed only after the end of the second World War, when research in radar and radio communications led to the development of the technical equipment needed for radio astronomy observations. However, even in the pre-war years, the well known Soviet physicists Academicians L. I. Mandel'shtam and N. D. Papaleksi had thought about the possibility of radar observations of the Moon.

Radio astronomy research in the Soviet Union began to be developed starting in 1946–1947. Thus, the 50th anniversary of the birth of radio astronomy essentially coincides with the 35th anniversary of its development in the USSR. In connection with this, the Scientific Council for Radio Astronomy of the Academy of Sciences of the USSR, headed by Academician V. A. Kotel'nikov, delegated a group of radio astronomers the task of preparing a publication outlining the history of the development of radio astronomy in the Soviet Union.

Radio astronomy is, naturally, a subfield of astronomy. Accordingly, one could lay out the history of its development following the division of astronomy into studies of various types of objects: radio astronomy of the solar system, of Galactic objects, of extragalactic objects, etc. However, because radio astronomy information about the Universe is obtained via measurements of radio emission arriving from the cosmos using radio physical techniques, it would also be reasonable to lay out such a history according to a different scheme: describing the history of radio telescope and radio receiver construction, the development of radio interferometry, spectral radio astronomy, radar astronomy, out-of-atmosphere radio astronomy, and so forth.

It would clearly be very difficult to write a monograph following either of these schemes within a rather compressed time schedule. In addition, it would be difficult to reflect in such a monograph various interesting details of the establishment and development of radio astronomy studies in various scientific groups across the country. Therefore, it was decided to first prepare a collection of essays, in which the pioneers of radio astronomy themselves or their close coworkers and students would describe the history of the development of radio astronomy research within their own institutes.

Of course, the writing of such a collection did not exclude, and more likely stimulated, the writing of various memoirs, such as the leaflet by Corresponding Member of the Academy of Sciences I. S. Shklovskii [1] or the article on Adacemician V. L. Ginzburg [2] published in 1982. The proposed collection was meant to, in part, supplement such previously published reviews, dedicated to the early period and results of the development of Soviet radio astronomy [3, 4, 5, 6].

When compiling the collection, the choice of a standard for citations proved to be a serious problem. The huge number of references for a book of limited volume precluded the inclusion of a comprehensive bibliography. We settled on a format in which the descriptions of the methods and results are accompanied only by an indication of the relevant authors and, as a rule, the publication date for the corresponding work. This supposes that the reader can find more detailed information using compilations of references, such as [7, 8].

It was considered necessary to allocate at least a little space in the collected essays to biographical information about the founders of Soviet radio astronomy who are deceased—N. D. Papaleksi, S. E. Khaikan, V. V. Vitkevich, G. G. Getmantsev, S. A. Kaplan, S. I. Syrovatskii, S. B. Pikel'ner, B. M. Chikhachev, M. M. Kobrin, D. B. Korol'kov.

When examining the history of the development of radio astronomy in the Soviet Union as a whole, we can mark several successive stages. The first began in 1946 with the theoretical work of V. L. Ginzburg and I. S. Shklovskii, radio observations of the solar eclipse of 1947 carried out by S. E. Khaikin and B. M. Chikhachev at the initiative of N. D. Papaleksi, the work of radio physicists in Gorkii under the supervision of M. T. Grekhovaya and G. S. Gorelik. This first stage continued until roughly the middle of the 1950s. At that time, primarily in relation to solving important applied problems having to do with the propagation of radio waves through the Earth's atmosphere, the first generation of Soviet radio telescopes, radio interferometers and radiometric receivers operating at a wide range of wavelengths from 4 m to 2 cm was created (partly based on radar technology), and the first series of radio observations of the quiescent and perturbed Sun, the Moon and the brightest discrete cosmic radio sources were carried out.

Astrophysical studies of fundamental importance were carried out in this stage, which opened possibilities for the development of a number of new directions in radio astronomy. As in all subsequent stages, a huge role in the development of Soviet radio astronomy was played by fundamental theoretical studies of the physics of the Sun, planets, interstellar medium, cosmic rays, supernova remnants, galactic nuclei, extragalactic objects and the expanding Universe (at the Sternberg Astronomical Institute of Moscow State University [GAISH], Lebedev Physical Institute of the USSR Academy of Sciences [FIAN], Institute of Applied Mathematics of the USSR Academy of Sciences [IPM], Byurakan Astronomical Observatory of the Academy of Sciences of the Armenian SSR [BAO], and the Radio Physical Research Institute in Gorkii [NIRFI]). These studies served as the main program for carrying out observations and interpreting the results obtained. Fundamentally new methods that are now in wide use in the world of radio astronomy were proposed in this period.

Among the important results of this first stage, we should note the discovery of the supercorona of the Sun by V. V. Vitkevich (FIAN), the discovery of the linear polarisation of the diffuse cosmic radiation (NIRFI), the detection of circular polarisation of the radio emission of active regions on the Sun (Main Astronomical Observatory of the USSR Academy of Sciences in Pulkovo [GAO]), and the detection and study of the spectra of thermal and non-thermal sources at centimetre wavelengths (FIAN, NIRFI, GAISH).

The second stage is marked by the construction of large-scale radio telescopes operating in various wavelength ranges (metre, centimetre and millimetre), especially for the needs of the radio astronomy community: the Large Pulkovo Radio Telescope; the fixed 31-m radio telescope of FIAN in the Crimea; the fully steerable, precise 22-m radio telescope of FIAN in Pushchino, and then the 22-m radio telescope of the Crimean Astrophysical Observatory [CrAO] in Simeiz; the DKR-1000 wide-band cross radio telescope of FIAN; the Large Radio Interferometer of the BAO and the compositive eight-element antenna of the Deep Space Communications Centre in the Crimea. Very important studies of the radio emission of the Moon were carried out in this stage, which made it possible to determine a number of characteristics of its surface layers using radio astronomy methods (NIRFI, FIAN), and to detect a previously unknown heat flux from the lunar core (NIRFI). Radio "images" of the Sun and Moon were obtained at centimetre and millimetre wavelengths, including the polarisation of the radiation (FIAN, GAO), and ejections of solar material with speeds exceeding the escape speed were detected (FIAN). The radio emission of Venus and some other planets was studied, and the presence of a high surface temperature on Venus was established (FIAN, GAO). Unique investigations of the fluxes and polarisations of discrete radio sources were conducted, in particular, of the Crab Nebula (FIAN, GAO, GAISH, BAO), and work that ultimately led to the discovery of the variability of extragalactic radio sources was begun (GAISH).

In this stage, methods for precise, absolute measurements were developed, and an accurate catalogue of the fluxes of powerful discrete sources was compiled (NIRFI). The spectra of many sources were measured over a broad range from centimetre to decametre wavelengths (GAO, FIAN, the Institute of Radio Physics and Electronics of the USSR Academy of Sciences [IRE], GAISH and others). Methods for measuring the brightness distributions of sources using scintillation observations and the "coverage" method were refined.

The results of these and other measurements made it possible to lay the basis for a number of theories, such as those describing the magnetic bremsstrahlung mechanism for the radio emission of solar active regions (FIAN, GAISH, GAO, NIRFI), the thermal radio emission of the Moon and planets (NIRFI), the synchrotron and thermal radio emission in the continuous spectra of cosmic sources (FIAN, GAISH, NIRFI) and the radio lines of atoms and molecules (GAISH).

The vigorous development of Soviet radar astronomy made it possible, not only to refine knowledge of the astronomical unit, which was exceptionally valuable for astronautics, but also to determine a number of important characteristics of the planets (IRE and others).

At the beginning of the 1960s, theoretical and experimental work related to a new and promising direction began—the search for extraterrestrial civilisations using radio astronomy methods (GAISH, NIRFI, Special Astrophysical Observatory [SAO], BAO).

In the middle of the 1960s (the third stage), specialised areas in radio astronomy were widely developed. Radio recombination lines of excited hydrogen were discovered (GAISH, FIAN, GAO) and studies in this direction were developed. The fundamentally new method of Very Long Baseline Interferometry was proposed (FIAN, GAISH, GAO), and work was begun on its realisation (GAISH, FIAN, NIRFI, the Space Research Institute [IKI], BAO, CrAO, GAO).

At the end of the 1960s and beginning of the 1970s, Soviet radio astronomy received a new impulse in its development in connection with the construction of large radio telescopes, including the unique RATAN-600 telescope (SAO), the T-shaped UTR-2 radio telescope (IRE) operating at decametre wavelengths, the Large Scanning Antenna operating at metre wavelengths (FIAN) and the RT-25X2 transit radio telescope operating at millimetre wavelengths (NIRFI). The Simeiz–Pushchino interferometer was devised on the basis of the two corresponding 22-m antennas, and this system was used to successfully carry out Soviet and international VLBI experiments enabling the realisation of extremely high resolution (IKI, CrAO, FIAN).

In these years, Soviet radio astronomy also went beyond the limits of the Earth's atmosphere. In addition to studies at the longest wavelengths begun earlier (GAISH, NIRFI), the first experiments with parabolic antennas operating at submillimetre (FIAN) and centimetre (IKI) wavelengths on board orbiting, manned stations were carried out. Radio astronomy measurements of the characteristics of the surfaces and atmospheres of the planets and of the Earth as a planet were conducted on board automated spacecraft (IKI, IRE, FIAN).

Observations with high spatial resolution enabled the discovery of radio granulation on the Sun (GAO, SAO, CrAO, NIRFI, the Institute of Applied Physics of the USSR Academy of Sciences [IPFAN]), and the measurement of the temperature of the moons of Jupiter (FIAN, SAO, CrAO). Studies of the brightness distributions and polarisations of discrete sources were continued over a wide range of wavelengths (SAO, BAO). A number of sources were observed with high frequency resolution in spectral lines of hydrogen, hydroxyl, formaldehyde and water vapour (SAO, FIAN, GAISH, IKI). The first spectral radio lines at millimetre wavelengths were detected—radio recombination lines of hydrogen H56α (FIAN), and later, very low-frequency spectral lines—recombination lines of carbon with principle quantum numbers up to 630 (IRE).

The absence of fluctuations of the cosmic background radiation was established with high accuracy (SAO), in disagreement with all theories of fragmentation. Measurements of the intensity of the cosmic background radiation were carried out at a number of wavelengths (NIRFI, FIAN). Thousands of new radio sources were discovered at centimetre and decametre wavelengths (GAISH, SAO, IRE).

During the third stage, wide-ranging studies of pulsars were conducted at centimetre, decimetre, metre, and decametre wavelengths (FIAN, IRE, IKI), which led to the discovery of new sources and established a number of important properties in the structure of pulsar pulses.

The construction of the 70-m radio telescope of the Deep Space Communications Centre began in the Crimea at the beginning of the 1980s, on which radio astronomy investigations at wavelengths right down to 8 mm were initiated.

In this collection of essays, we can only make reference to all of the studies listed above.

In the 1950s, the Commission for Radio Astronomy was established in the Scientific Council of the USSR Academy of Sciences on Astronomy, under the chairmanship of S. E. Khaikin. Later (in 1961), this was transformed into the Scientific Council of the USSR Academy of Sciences on Radio Astronomy, whose permanent chairman is Academic V. A. Kotel'nikov. The Scientific Council on Radio Astronomy played and continues to play an extremely important role as a force unifying Soviet radio astronomers, in helping to realise large-scale scientific and technical projects, and in organising regular conferences that attract a large number of specialists in both radio astronomy and related fields. With the help of this Council, international collaborations have been developed between Soviet radio astronomers and their foreign colleagues, in which virtually every scientific institution working on radio astronomy problems in the Soviet Union participates.

Returning to the structure of this collection, we note that the arrangement of the contributions is fairly arbitrary. It is determined primarily by the time when radio astronomy studies began at the institutions involved. In recent years, both fundamental and applied radio astronomy have also developed in some organisations that are not directed represented in this collection. These include, for example, the Physical–Technical Institute of the Academy of Sciences of the TSSR, the Main Astronomical Observatory of the Academy of Sciences of the Ukrainian SSR, the Bauman Moscow Higher Technical Institute and some others. Works by researchers at these organisations are referenced in the corresponding contributions.

Radar studies of meteors in the USSR began in 1946 [4], and soon became a method for studying atmospheres. This method is not considered here.

In connection with the limited volume available, and also to avoid repetition in the material presented by the various scientific groups, the collection includes cross-references to other essays in the volume. For convenience in referencing, arbitrary identifying numbers are given in the headings to the essays.

The Editorial Group is grateful to all who took part in the creation of this collection, especially L. I. Matveenko and P. D. Kalachev, who looked over the manuscript and made valuable comments. We will also be thankful to any readers for comments that arise as they become acquainted with this collection.

S.Y. Braude
B.A. Dubinskii
N.L. Kaidanovskii
N.S. Kardashev
M.M. Kobrin
A.D. Kuzmin
A.P. Molchanov
Y.N. Pariiskii
O.N. Rzhiga
A.E. Salomonovich
V.A. Samanian
I.S. Shklovskii
R.L. Sorochenko
V.S. Troitskii
K.I. Kellermann

Contents

1 **Radio Astronomy Studies at the Lebedev Physical Institute** 1
B.A. Dogel', Y.P. Ilyasov, N.L. Kaidanovskii, Y.L. Kokurin,
A.D. Kuz'min, A.E. Salomonovich, R.L. Sorochenko, and V.A. Udal'tsov

2 **Radio Astronomy Studies in Gorkii** 61
A.G. Kislyakov, V.A. Razin, V.S. Troitskii, and N.M. Tseitlin

3 **The Development of Radio Astronomy at the Sternberg
Astronomical Institute of Lomonosov Moscow State University
and the Space Research Institute of the USSR Academy of Sciences** . 89
L.M. Gindilis

4 **The Department of Radio Astronomy of the Main Astronomical
Observatory of the USSR Academy of Sciences** 117
N.L. Kaidanovskii

5 **Radio Astronomy at the Special Astrophysical Observatory
of the USSR Academy of Sciences** 137
N.L. Kaidanovskii

6 **Radio Astronomy at the Byurakan Astrophysical Observatory,
the Institute of Radio Physics and Electronics of the Academy
of Sciences of the Armenian SSR and Other Armenian Organisations** 157
V.A. Sanamian

7 **The Development of Radio Astronomy at the Crimean
Astrophysical Observatory of the USSR Academy of Sciences** 171
I.G. Moiseev

8 **The Development of Radio Astronomy Research at the Institute
of Radio Physics and Electronics of the Academy of Sciences
of the Ukrainian SSR** 181
S.Y. Braude and A.V. Megn

9 **Radio Physical Studies of Planets and the Earth at the Institute of Radio Technology and Electronics of the USSR Academy of Sciences** 205
B.G. Kutuza and O.N. Rzhiga

10 **Radio Astronomy Studies of the Sun at the Institute of Terrestrial Magnetism, the Ionosphere and Radio-Wave Propagation of the USSR Academy of Sciences** 225
S.T. Akinian, E.I. Mogilevskii, and V.V. Fomichev

11 **The Birth and Development of Radio Astronomy Studies of the Sun at the Siberian Institute of Terrestrial Magnetism, the Ionosphere and Radio-Wave Propagation** 231
G.Y. Smol'kov

12 **Radio Astronomy at the Radio Astrophysical Observatory of the Latvian SSR Academy of Sciences** 237
A.E. Balklavs

13 **Radio Astronomy Research at Leningrad State University** 243
A.P. Molchanov

Index 247

Chapter 1
Radio Astronomy Studies at the Lebedev Physical Institute

B.A. Dogel', Y.P. Ilyasov, N.L. Kaidanovskii, Y.L. Kokurin, A.D. Kuz'min,
A.E. Salomonovich, R.L. Sorochenko, and V.A. Udal'tsov

Abstract The history of the development of radio astronomy studies at FIAN is described, beginning with the first theoretical (1946) and experimental (1947) studies of the solar radio emission. Information about the development of the Crimean station of FIAN, then the establishment and development of the Radio Astronomy Station in Pushchino is presented. Work on the construction of large radio telescopes, including the FIAN 22-m, DKR-1000 and BSA telescopes, is described, together with important results obtained during observations of the Sun (including the discovery of its "supercorona"), planets, line radio emission and studies of pulsars and other discrete sources.

1.1 The First Steps[1]

The important radio physicist Academician Nikolai Dmitrievich Papaleksi (Fig. 1.1) is justifiably considered the founder of radio astronomy research in the Soviet Union. Nikolai Dmitrievich was interested in astronomy and meteorology even in his youth. Before the Second World War, he and Academician Leonid Isaakovich Mandel'shtam considered the possibility of measuring the distance to the Moon using radar methods, by detecting the time delay of a radio pulse sent from the Earth and reflected off the lunar surface. The level of radio technology available at that time (1925) made this appear unpromising. Papaleksi and Mandel'shtam returned to this problem during the years of the Second World War. Their new calculations, published at the beginning of 1946, showed that such measurements were realistic, and indeed, they were carried out in that same year in Hungary and the USA. When it came to thinking of radar measurements of the Sun, Papaleksi gave the young theoretician of the Lebedev Physical Institute (FIAN) V. L. Ginzburg the problem of carrying out the necessary calculations.

---

[1]Section 1.1 was written by N.L. Kaidanovskii and A.E. Salomonovich, Sect. 1.2 by V.A. Udal'tsov, Y.L. Kokurin and R.L. Sorochenko, Sect. 1.3 by Y.P. Ilyasov, A.E. Salomonovich and A.D. Kuz'min, Sect. 1.4 by V.A. Dogel' and Sect. 1.5 by A.E. Salomonovich.

Fig. 1.1 Nikolai Dmitrievich Papaleksi (1880–1947)

Academician Ginzburg describes this occurrence as follows: "N. D. Papaleksi, naturally, had thought about radar measurements of the planets and Sun. In this connection, he asked me at the end of 1945, or more likely the beginning of 1946, to elucidate the conditions for the reflection of radio waves from the Sun. It stands to reason that, in essence, this was a typical ionospheric problem, and I had all the corresponding formulas to hand. The results of the calculations did not seem especially optimistic, since they indicated that, for a broad range of parameters, many of which were unknown then (the number density of electrons, the temperatures in the corona and chromosphere), radio waves should be strongly absorbed in the corona or chromosphere, so that they should not even reach the level where they would be reflected... But a more interesting conclusion followed directly from this: the sources of solar radio emission should not be in the photosphere, but instead in the chromosphere, or even in the corona in the case of longer waves. Further, it was already supposed at that time that the corona was heated to hundreds of thousands, or even a million, degrees. Thus, even under equilibrium conditions (in other words, in the absence of any perturbations), the temperature of the solar radio emission emitted by the corona (waves with wavelengths longer than about a metre) should reach about a million degrees for a photospheric temperature of 6000 degrees" [2, p. 289].

These results were laid out by Ginzburg in a paper published in the Reports of the Academy of Sciences in 1946. The conclusion that the corona must be the source of solar radio waves at metre wavelengths was also drawn nearly simultaneously and independently by I. S. Shklovskii in the Soviet Union and by D. Martin in England.

Papaleksi had also been interested earlier in the problem of solar–terrestrial connections, including the important question of the influence of solar activity on the Earth's ionosphere and the stability of radio communications. One of the methods he adopted was making observations during solar eclipses, when it was possible to distinguish the influences of the photon and particle fluxes from the Sun. During the eclipse of July 9, 1945, Papaleksi had already carried out a broad set of studies of phenomena in the ionosphere accompanying the eclipse. He intended to continue these investigations during the total solar eclipse of May 20, 1947, which it was possible to observe from Brazil. For this purpose, he began to prepare a large, multi-faceted expedition. Now, after the estimates of Ginzburg, it was planned for the first time to make not only ionospheric observations, but also direct observations of the radio emission of the Sun at metre wavelengths. Papaleksi hoped to detect not only the steady-state emission of the quiescent Sun, but also (thanks to the high resolution attained during eclipse observations) regions of sporadic radio emission. Information about this radiation obtained abroad during the war years was just starting to appear in literature at that time.

In a public lecture in January 1947, Papaleksi said the following about radio astronomy: "This new area of research, which is currently in its infancy, will undoubtedly be of extreme interest for physics of the Sun. There is every reason to believe that the application of radio astronomy methods in astronomy will open a new era, whose importance can be compared to the discovery of Fraunhofer lines and the application of spectral analysis in astrophysics, and which will help us penetrate more deeply into the mysteries of the Universe."

Radio observations of the Sun during partial eclipses began abroad starting in 1945, but these observations did not yield conclusive results. The total eclipse of May 20, 1947 could potentially provide important new information.

The Brazilian expedition was organised by the Scientific Council of the USSR Academy of Sciences on Astronomy under the supervision of Academician Papaleksi, who was then the Head of the Oscillation Laboratory of FIAN. The well known polar explorer G. A. Ushakov was the administrative assistant to the head of the expedition. The expedition included researchers from optical, radio astronomy and ionospheric groups. Papaleksi supervised the last two of these groups, and M. N. Gnevyshev the optical group.

In the period of preparation for the expedition, it was necessary to acquire and adapt all the equipment, develop the methods to be used for the observations, carry out the necessary calculations of the conditions for the eclipse and prepare a program for the reduction of the observations.

The expedition members included the now well known scientists V. L. Ginzburg and I. S. Shklovskii. The radio astronomy observations were to be carried out by Papaleksi's student B. M. Chikhachev (Fig. 1.2), formerly of the Central Radio Laboratory, who was an experienced specialist on radio technology and electronics. Before the Second World War, he had studied the technology of preparing metallic radio lamps in the USA, and, during the war, he was the chief technician at a radio factory. After the war, he became a PhD student of Papaleksi in FIAN. Workers under Academician A. I. Berg were assigned as assistants to Chikhachev. This group was responsible for the radio astronomy equipment for the expedition.

The planned observations of the radio emission of the Sun at a wavelength of 1.5 m required a radio telescope with a fairly high sensitivity, since, according to calculations, the intensity of the signal might be decreased by more than a factor of ten during the eclipse. An antenna with a large receiving area that was capable of tracking the Sun over the entire period of the eclipse (about three hours) was required, as well as a broad-band receiver with a good noise coefficient. A large

Fig. 1.2 Boris Mikhailovich Chikhachev (1910–1971)

number of Soviet and foreign radar equipment with suitable parameters were left after the war. A radar receiver operating at a wavelength of 1.5 m was selected for the eclipse observations.

However, Papaleksi was not able to realise his plan; he died suddenly on February 3, 1947. At a memorial service dedicated to N. D. Papaleksi in April 1947, the President of the Academy of Sciences Academician S. I. Vavilov said, "Death has claimed Nikolai Dmitrievich during his preparations for an expedition to Brazil, for which he and his students, as always, prepared meticulously and at the head of which he stood. He was not fated to live to see the realisation of this expedition, and now his orphaned students will sail on a Soviet ship to the shores of Brazil without their teacher.[2]"

The supervision of the expedition was given to Corresponding Member of the Academy of Sciences A. A. Mikhailov, and his scientific assistant and supervisor of the radio astronomy and ionospheric groups became Semen Emmanuilovich Khaikin (Fig. 1.3)—a student of L. I. Mandel'shtam, who can be considered a co-founder of Soviet experimental radio astronomy. At that time, he was the head of a group in the Oscillation Laboratory of FIAN, and was working on studies of the physical properties of solid bodies and electrolytes using radio methods.

The experience acquired by Khaikin during the Second World War with the construction of radar systems helped him develop a method for observing the radio emission of the Sun during an eclipse.

The expedition began with an unpleasant surprise; a shelf of heavy ice with a width of 5 km formed in the usually unfrozen port of Liepaya from which the ship *Griboedov* (Fig. 1.4) was to set sail for Brazil. The icebreaker *Sibiryakov* was called from Riga, and was supposed to make a passage through the ice in order to lead the ship to clear water. After the war, there were many mines left in the port, and there was a real danger of getting blown up by one. The *Griboedov* arrived in Sweden only

---

[2]Izvestiya Akademii Nauki SSSR, Seriya fizika, 1948, 12, No. 1, p. 5.

Fig. 1.3 Semen Emmanuilovich Khaikin (1901–1968)

Fig. 1.4 Participants of the Brazilian expedition of the USSR Academy of Sciences on the desk of the "Griboedov". Included in the picture are the well known scientists and engineers S. E. Khaikin (*bottom row, first on the right*), G. A. Ushakov (*lower row, fourth from the right*), V. L. Ginzburg (*middle row, fourth from the left*), B. M. Chikhachev (*middle row, sixth from the right*), and I. S. Shklovskii (*upper row, second from the right*)

on April 15, where it underwent demagnetisation over the course of two days. The ship arrived at the Brazilian port of Salvador on May 10, 1947, only ten days before the eclipse. The optical and ionospheric groups set out by train to the city of Arasha.

There was now little time to set up and test the equipment. In spite of its characteristics, which were quite good for that time, the radar station was not usable for observations of solar radio emission during the eclipse without certain adaptations. When mounted in the usual way, the antenna had a vertical axis that could rotate in azimuth, and was not able to carry out observations at large elevations—and the centre of the Sun during the eclipse would be at elevations of approximately 35 to 57°.

Khaikin solved this problem in the following very clever way: the surface of the antenna was laid on the deck and inclined about its horizontal axis using winches. Rotation in azimuth was achieved by rotating the ship itself. Since the measurements were carried out in a bay whose bottom would not take a secure hold on the anchor, lines for rotating the vessel were strung to the shore. This provided tracking of the Sun during the observations with an accuracy of about 2°—quite sufficient for this purpose.

The receiver was calibrated using the intensity of its own noise, and was estimated in units of the maximum noise with an accuracy of 5%. The output electromagnetic recorder worked poorly, and there weren't chart recorders at that time. And of course, the results of the observations must be absolutely trustworthy. At Khaikin's suggestion, output pointer-display instruments were added in parallel, and placed in isolated cabins where observers could independently write the readings from these instruments each minute, using clocks checked against the ship's chronometer. The position of the ship was also noted each minute. Corrections for the deviation of the antenna and monitoring of the noise correction were carried out every five minutes.

The emission received could include a contribution from the Galaxy as well as the Sun. Therefore, measurements of the radio emission from the same region of the sky were carried out 15 days later, when the Sun was offset by 15° and was outside the beam of the antenna. It turned out that Galactic emission comprised only about 5% of the receiver noise, in other words, it was at the sensitivity limit, so that no correction to the eclipse curve was needed. The eclipse curve (Fig. 1.5) showed that the minimum intensity of the solar radio emission was only 40% lower than the maximum intensity—the eclipse was annular, or ring-like, rather than total for radio waves with 1.5 m wavelength. The intensity was higher in the first than in the last phase of the eclipse, indicating that the distribution of the solar radio brightness was not uniform. If we suppose that the radio brightness distribution over the solar disk was uniform, the radiating layers of the Sun should reach distances from the solar centre of about 0.35 times the radius of the Sun ($\sim 0.35 R_\odot$) above the photosphere. Since this was clearly not correct, the derived radio diameter of $2.7 R_\odot$ was only approximate. Nonetheless, this value was in good agreement with the theoretical calculations of Ginzburg.

The brightness distribution derived from the observations agreed with the arrangement of spots on the Sun. An eclipse curve constructed by E. I. Mogilevskii based on data for protuberances and filaments from a number of observatories proved to be close to the observed radio curve.

The results of these solar-eclipse observations carried out under Khaikin became classic—the theoretically predicted coronal radio emission at metre wavelengths had been detected. Some 23 years after the observations (on April 28, 1970, with the discovery date given as October 28, 1947), these results were noted with a diploma

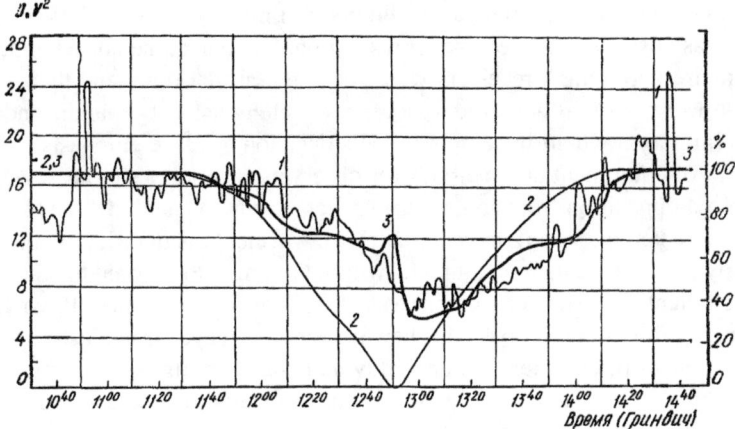

Fig. 1.5 Recording of the radio eclipse of the Sun obtained by Soviet radio astronomers on May 20, 1947: (**1**) variation of the intensity of the solar radio emission at 1.5 m wavelength in arbitrary units; (**2**) variation of the visible area of the solar disk; (**3**) behaviour of the "eclipse" of eruptive prominances and filaments. The horizontal axis denotes GMT

honouring the discovery, in the names of N. D. Papaleksi, S. E. Khaikin and B. M. Chikhachev.

The observations of the solar radio emission during the eclipse so enthralled Khaikin that he decided to "pack up" his earlier work and devote all his energy to radio astronomy. Upon his return from the Brazilian expedition, he composed a broad plan for the development of this area in the Soviet Union—not an easy task. None of the required measurement techniques were available, and the country had no specialists whose work was close to the relevant astrophysical questions or technical problems in statistical radio physics. Without interest in these problems, it would not be possible to make plans for radio astronomy observations or to develop the needed methods.

Khaikin suggested that, although radio astronomy was above all an astronomical science, in that early time, when the required experimental methods were first being developed, the most appropriate workers for this field were radio physicists, and the most appropriate bases for their work were physics institutes. He believed that, as radio astronomy grew toward becoming a true branch of astronomy, these first groups would be supplemented with astrophysicists, and bases for radio astronomy would appear among astrophysical observatories as well as physics institutes.

Khaikin also understood that, although the development of radio astronomy receiver technology—radiometers—was a progressing area, possibilities for increasing the sensitivity of receivers were limited, while there were virtually no limitations on increasing the area or resolution of antennas. Therefore, the development of radio astronomy technology required first and foremost antennas with very large areas—preferably reflectors and not refractors (co-phased antennas).

Khaikin correctly predicted the progressive tendency in radio astronomy to move toward ever shorter wavelengths—decimetre and centimetre wavelengths, where the noise temperatures of the antennas would not be limited by the Galactic noise, and where it was possible to achieve better resolution. Of course, the noise temperatures of centimetre-wavelength receivers were very high at that time, and their sensitivity was lower than that achieved at longer wavelengths; but Khaikin understood that this limitation would be temporary. His decision to place emphasis on the development of equipment and methods for observations at ever shorter wavelengths encountered opposition from some astrophysicists, who wanted to stick to observations using radar equipment operating at metre wavelengths in order to obtain radio astronomy measurements more quickly, without taking into account the fact that the scientific potential of such equipment was already nearly exhausted. It would be an exceedingly difficult task to base this new branch of science—radio astronomy—on the development of complex and unwieldy antennas and high-sensitivity receivers "in a vacuum"—without any technological basis or qualified scientific researchers or service personnel, and without the corresponding financing. Therefore, Khaikin proposed a research programme concerning the propagation of radio waves with wavelengths from 3 m to 3 cm through the entire thickness of the Earth's atmosphere, using the Sun, Moon and discrete radio sources as generators of radio emission beyond the atmosphere.

In addition to the refraction of radio waves arriving at various elevations in the troposphere and ionosphere, the attenuation of the signals due to absorption and scattering were measured. Along the way, the fluxes, angular dimensions and precise coordinates of the various sources of radiation were to be determined. Such studies were carried out under the supervision of Khaikin, as well as by a group of radio astrophysicists at Gorkii State University headed by Prof. G. S. Gorelik. The advantage of such a programme, sitting at the cross-roads of radio physics and astronomy, was that it enabled physicists working in the field to acquire knowledge in astronomy and to develop intuition about radio astronomy, and also enabled refinement of the technological resources available. This made it possible for this work to grow into true experimental radio astronomy, encompassing both astrophysical and astrometric problems.

In 1948, at the suggestion of the Presidium of the USSR Academy of Sciences, a programme of this work was adopted for 1948–1950. This programme foresaw the transfer of a number of radar stations operating at metre, decimetre and centimetre wavelengths to FIAN and Gorkii State University, together with the required funding. Engineering and technical staff and corresponding funding were also allocated for affiliate stations in other locations.

We should emphasise that, from the very beginning, the Director of FIAN and President of the Academy of Sciences Academician S. I. Vavilov solidly supported Khaikin's initiative, consistently supporting radio astronomers in the future as well. The same is true of Academician M. A. Leontovich—the new head of the Oscillation Laboratory.

After this decision to develop radio astronomy methods for studies of radio-wave propagation, the Crimean Station of FIAN was reorganised. A former school on Sevastopol' Road in Alupka became the central base, where the administrative headquarters, a workshop and a small dormitory were located. In addition to this centre,

Fig. 1.6 Viktor Vitol'dovich Vitkevich (1917–1972); photo taken in 1948

the Crimean Station had the Alushta base, which was located on a parcel of seashore in Rabochii Ugolok, where a Finnish house with a basement for a workshop and a stone shed with a diesel generator was built in 1949. (Both locations were initially intended for the continuation of pre-war work by Papaleksi on interferometric distance measurements.) B. N. Gorozhankin of the FIAN Oscillation Laboratory was assigned as the Head of the Crimean Station.

It was clear that the sites in Alupka and Alushta were insufficient for work on a large scale. In agreement with the director of the Crimean Astrophysical Observatory Academician G. A. Shain, the Crimean Station of FIAN was given a site (250×150 m) on Koshka Mountain near Simeiz, bordering on the observatory, intended for large antennas, laboratory buildings, a workshop, a storeroom and a dormitory.

Khaikin attracted V. V. Vitkevich (Fig. 1.6) to this new work in 1948. He had finished postgraduate studies at Moscow State University by correspondence in 1944, being at that time an officer in the Navy, and had been working toward his doctorate in the Radio Technology Section of the Academy of Sciences since 1947. An experienced radio specialist and senior scientist, Vitkevich headed a group that soon also included the engineers D. V. Kovalevskii and V. S. Medvedev, and later the radio technicians M. T. Levchenko, L. A. Levchenko, E. K. Kurkina (née Karlova), T. I. Shakhanova (née Gavrilenko) and others. This group set about the work of devising radiometer and radio-interferometer equipment operating at decimetre wavelengths. In the Spring of 1949, successful tests were carried out of an interferometer operating at 50-cm wavelength comprised of 3-m antennas from a radar system and a receiver supplemented by a modulating radiometer, in which Vitkevich used a capacitor switch he had developed. However, the volume of work needed required the application of considerable efforts.

Various co-workers of Khaikin (Ya. I. Likhter, N. L. Kaidanovskii and A. E. Salomonovich) who had returned after demobilisation in 1946 and continued work on their pre-war topics—radio physical studies of solids and liquids—turned their attention to the new research direction that was being developed. In place of B. N.

Gorozhankin, who headed the Group for the Mathematical Processing of Observations (L. N. Borodovskaya, V. M. Antonova), Salomonovich was appointed as the Head of the Crimean Station, Kaidanovskii as his assistant, Vitkevich as the Head of the Alushta operation and the engineer Kovalevskii as the Head of the operation on Koshka Mountain. This reorganisation made it possible to speed up the work being carried out. Fourth-year (R. L. Sorochenko, F. V. Bunkin, B. D. Osipov, N. B. Delone) and third-year (V. V. Kobelev, N. V. Karlov, T. A. Shaonov, V. G. Veselago) students from the recently founded Physical–Technical Institute were also assigned to the Crimean Station, and the graduates of the Moscow Energy Institute N. F. Ryzhkov and T. M. Egorova were taken on in 1950.

The development of measurement methods was carried out under the overall supervision of Khaikin, at metre wavelengths by Chikhachev and Likhter, at decimetre and then metre wavelengths by Vitkevich, at 10 cm by Kovalevskii and Sorochenko and at 3 cm by Kaidanovskii. Measurements of refraction at metre and decimetre wavelengths were conducted using a so-called sea radio interferometer, which used the sea surface to reflect the radio waves—a radio analog of a Lloyd reflector.

The use of the sea interferometer made it possible to measure refraction in a relatively simple way at metre and long decimetre wavelengths, with good accuracy based on the times of passage of the Sun through the lobes of a multi-lobe antenna beam. The roughness of the sea (whose surface served as the Lloyd reflector) could be neglected for observations near the horizon. By that time, a number of powerful discrete sources with small angular sizes were known, enabling measurements using comparatively narrow lobes of the interferometer beam.

An antenna operating at 1.5 m was installed on the site at Koshka Mountain at a fairly high altitude above sea level ($\sim$285 m) in 1948 under the supervision of Gorozhankin. The receiver had an attachment at its output that enabled correction for the noise of the receiver itself.

In the Summer of 1949, Vitkevich and his group set about using the sea interferometer at Alushta to take measurements of the solar radio emission at 50-cm wavelength, which did not require high altitudes above sea level. The antenna was mounted on an angular turret pointed at the Finnish house, at a height of several metres above sea level. Later, in 1950, two-antenna radio interferometers were constructed at the cape of Ai-Todor. The source Taurus A (in the Crab Nebula) was detected for the first time in the Soviet Union using one of these interferometers operating at 1.5 m wavelength (Vitkevich, Rykhkov, Medvedeva). Measurements with interferometers at high altitudes were carried out in 1949–1952, on the mountains Ai-Petri (Osipov, Delone, Kurkina) and Kastel' (Rykhkov and others).

When taking measurements of the Sun at centimetre wavelengths, it was possible to get by without interferometric observations, using parabolic antennas with diameters corresponding to about 100 wavelengths. The angle of refraction could be determined by tracking the Sun in azimuth and noting the times when its centre passed through the axis of the vertical beam of the antenna, which was mounted at specified angles. Thus, radio telescopes suitable for observations at wavelengths of 3 and 10 cm were sufficient for this work.

A radio telescope with a diameter of 7–10 m was needed for observations at 10 cm wavelength. The mount of the 7.5-m diameter parabolic dish from a

Wurzburg–Riese German radar system operating at 50 cm wavelength—a trophy of the war—was used in its construction. Its turning apparatus, found in another place, had to be rebuilt and augmented. In addition, the 10-cm observations required a new, more accurate reflecting surface of solid sheet aluminium. The work on this radio telescope was headed by the highly qualified construction engineer and Doctor of technical sciences P. D. Kalachev, who joined FIAN in 1948, and has done much for Soviet radio astronomy. The assembly of the mechanisms and constructional elements was carried out by N. S. Bulanov and P. I. Maiorov, under the supervision of P. M. Butuzov. The development of the technology for preparing the reflecting surface was done by Kalachev, Kaidanovskii and Salomonovich. This problem was solved using a specialised template and concrete matrix to punch out the reflector elements. In May 1949, all the components of the radio telescope had been sent from Moscow to Koshka Mountain, where the assembly of the telescope began immediately.

The constructional rigging for the large-scale and heavy components of the radio telescope was strung in the most primitive way, using two parallel block-and-tackle setups based on tubular tripods. Fortunately, the entire operation was carried out without incident. S. I. Kharlamov punched out the aluminium sheets on the concrete matrices with an accuracy to within 1 mm. However, it was not possible to attach these sheets to the crude frame, originally intended for observations at 50 cm, since this would degrade the accuracy of the reflecting surface. Therefore, well dried pine lathes were fixed to the ribs of the frame, which were then carefully planed, leaving no gaps, placed beneath a "flag" and painted. The punched-out aluminium sheets were then screwed onto the pine lathes. The shape of the resulting surface was monitored with accuracy to 1 mm using the "flag."

The prepared reflecting surface was carried to the telescope mount by hand using a specialised platform, raised and fixed to the mount. All the assembly and mechanical adjustment for the 7.5-m diameter radio telescope were finished in July 1949, after which the process of outfitting the telescope with the necessary radio and technical equipment began under the supervision of Kovalevskii and Sorochenko. This was how the construction and outfitting of the RT-7.5—the first Soviet reflector radio telescope—was brought about (Fig. 1.7). After it was equipped with a radiometer operating at 10 cm wavelength, this radio telescope was long used for observations of the Sun and Moon. It provided data on the propagation of radio waves at this wavelength through the Earth's atmosphere, and was later used to conduct a series of measurements of the radio emission of the Sun, Moon and other sources.[3]

The 3-m diameter antenna of a "Small Wurzburg" radar system, originally operated at 50 cm wavelength, was used in the construction of a radio telescope for observations at 3 cm wavelength. The surface was aligned by Kharlamov to an accuracy sufficient for such observations. The radio telescope was set up on the Alushta

---

[3] Many years later, a researcher in the Crimean station, Yu. I. Alekseev, used this radio telescope for polarisation measurements of the Sun. The reflecting surface was replaced by a co-phased antenna operating at 1.64 m, but the RT-7.5 mount continued to be used for this new purpose.

Fig. 1.7 First large reflecting radio telescope of the Crimean station of the FIAN, with a diameter of 7.5 m

site, right by the seashore. The Small Wurzburg radar system was augmented with radio equipment and the required measurement devices.

A Dicke disk modulator and absorbing valve were inserted into the antenna tract, to be used as an attenuator during the calibration of the radiometer when the radio telescope was pointed at the zenith. A high-frequency block with a heterodyne and preliminary cascade of intermediate-frequency amplifiers were mounted in the antenna. The subsequent cascades were installed in the laboratory. The apparatus for the 7.5-m diameter radio telescope operating at 10 cm was implemented in the same way.

The first metre-wavelength radio telescopes, used first for refraction measurements and then also for astronomical observations, were multi-dipole co-phased antennas set to receive at a fixed wavelength and usually outfitted with compensating radiometer receivers. Examples include the radio telescope operating at 1.5 m wavelength described above, later improved by Kalachev to enable variation of the elevation of the electrical axis via mechanical rotation about the horizontal axis. Antennas operating at 2 and 4 m wavelength were likewise constructed or adapted with record speed.

In June–July 1949, all the observational installations were ready, and systematic measurements of the refraction and attenuation of signals as a function of the elevation and observing conditions were begun. These observations continued for about a year. In 1950, the brightness temperature of the Sun was estimated at various wavelengths.

Simultaneous with the observations at the Crimean Station, similar measurements were carried out at the Gorkii Physical–Technical Institute by a large group under the supervision of G. S. Gorelik and V. S. Troitskii. A general report on this work, confirmed by Vavilov, was presented on June 15, 1950, and members of this group were awarded the prize of the Council of Ministers of the USSR.

In this period, experience in solving methodological problems accumulated, experimental measurement techniques were developed and sensitive receivers capable of operating over a wide wavelength range were devised. In addition, the physicists and engineers involved acquired some experience with astronomical observations and interest in astrophysical problems. Technological possibilities for improving the methods used to measure various physical characteristics of cosmic radio sources appeared. Beginning in 1949, at the initiative of Khaikin, Vitkevich and Chikhachev, a group of constructors headed by Kalachev began to plan and then construct new radio telescopes intended primarily for radio astronomy observations. These second-generation radio telescopes are described below.

Chikhachev carried out broad and interesting studies of active regions on the Sun at metre wavelengths, which became the basis for his PhD thesis—the first dissertation on radio astronomy at FIAN and only the second in the Soviet Union[4] Vitkevich was also undertaking fruitful activity in radio astronomy at that time, and put forth in 1951 the idea of studying the solar corona by analysing the radiation from a background source (the Crab Nebula) that had passed through the coronal material.

Studies of the propagation of radio waves through the Earth's atmosphere using radio astronomy methods were continued in 1951–1953, with the Radio Physical Research Institute in Gorkii also taking part. The scientific supervisor for this new series of studies was Khaikin, while the supervisors of individual areas of research were Kaidanovskii, M. A. Kolosov, Troitskii and Chikhachev. In addition to the observations in the Crimea and Gorkii, systematic measurements of the Sun and Moon were carried out at the newly organised station in the Kaluga region.

The radio emission of the Sun was used earlier for propagation measurements at centimetre wavelengths, with the measurements carried out when the Sun was rising or setting. It was also important to know the run of the measured quantitites throughout the day, which could be derived from observations of the Moon, which was the second most powerful source of centimetre-wavelength radiation after the Sun. Work in this area was led by the group of Kaidanovskii. The sensitivity of the radio telescope operating at 3 cm was substantially enhanced, and the diameter of the parabolic antenna was increased to 4 m.

The upgrade of this antenna by increasing the diameter to 4 m and the radiation angle to 120° was carried out under the supervision of Kalachev. The sensitivity

---

[4]The first was the thesis of V. S. Troitskii. We do not indicate here all subsequent dissertations in this field, and explicitly make note only of these two theses.

of the radio telescope was 1–2 K, with a time constant of one second. The radio technicians N. A. Amenitskii, A. A. Gromadin and S. K. Palamarchuk and the technical assistants V. P. Izvarina and E. P. Morozova participated in the construction and exploitation of this equipment.

To investigate the influence of climatic and meteorological conditions, one of the new 3-cm wavelength radio telescopes was installed in the Crimea at Koshka Mountain and another in the Kaluga region. Joint measurements with these telescopes were fully completed in December 1952.

In parallel with the radio-physical measurements at the Kaluga and Crimean Stations, measurements of the brightness temperature of the Moon at wavelengths of 3 and 10 cm as a function of the lunar phase were initiated. Measurements of the brightness temperature of the Moon during eclipse were also carried out, which did not show any changes.

During the course of these investigations, M. T. Turusbekov first looked for inhomogeneity in the brightness temperature across the lunar surface by looking for shifts of the "centre of gravity" of the emission. No such shifts were detected over the duration of the series of observations with an accuracy to within 0.5 min, in agreement with calculations based on the theory of Troitskii, which indicated that possible variations in the mean temperature of the Moon should be no larger than 7–10 K.

At the Kaluga Station, the 3-cm receiver was fully rebuilt in preparation for measurements of the polarisation of the solar radio emission in 1953–1954.

The original methods and equipment developed by E. G. Mirzabekyan (1912–1979) under the supervision of Khaikin and Kaidanovskii subsequently led to interesting astrophysical results (see Essay 4). In October–December 1954, using this same radiometer in a mode in which unpolarised radiation was received, Kaidanovskii and N. S. Kardashev (then a Master's student at Moscow State University) conducted successful observations of three supernova remnants and, for the first time at this wavelength, the diffuse Omega and Orion nebulae. These observations confirmed the theoretical conclusions of Shklovskii. By the end of 1954, all the FIAN radio telescopes were freed of their applied tasks and were fully turned toward astrophysical observations.

In this way, the plan of Khaikin was gradually realised: from applied tasks that could be solved using rebuilt radar equipment that was already available, to radio astronomy observations using radio telescopes constructed especially for this purpose.

Back in 1952, in contrast to Khaikin and Kaidanovskii, most prominent coworkers in the Radio Astronomy Section of FIAN thought it would be expedient to move to astronomical observatories. However, it proved to be more efficient to carry out this work within the walls of a multi-department physics institute. Unfortunately, there was also a certain attitude of "we don't need astronomers, except perhaps to calculate coordinates." This led to the loss from FIAN of a number of talented graduates of Moscow State University, including the astrophysicists Yu. N. Pariiskii and N. S. Soboleva, who carried out their Master's work in the Radio Astronomy Section of FIAN (under the supervision of Salomonovich).

Before 1954, in contrast to the Byurakan and Crimean Observatories (see Essays 6 and 7), the Pulkovo Observatory, which was the largest in the Soviet Union, did not yet have a radio astronomy department, and Khaikin took on the organisation of such a department. The subsequent development of radio astronomy at FIAN was associated with work at the Crimean Station under the supervision of Vitkevich, and with preparations for and then the construction of the new Radio Astronomy Station in Pushchino. After Khaikin stepped down as head of the Radio Astronomy Section of FIAN in 1955 in connection with his final move to the Pulkovo Observatory, this position was occupied by Vitkevich, assisted by Salomonovich.

1.2 The Crimean Station of the Lebedev Physical Institute (1952–1962)

Using the existing experimental base of the Crimean Station and throughout its continuous upgrades, radio astronomy investigations of the Sun, Crab Nebula, Galaxy (at 21 cm) and ionosphere were widely expanded. Possibilities for the further development of radio astronomy studies at the Crimean Station were essentially exhausted in this stage.

The head of the Crimean station in 1952 was Vitkevich and then, beginning in 1958, Yu. L. Kokurin. The assistant head for scientific and technical matters was Kovalevskii from 1950–1954, followed by V. A. Udal'tsov from 1954–1958.

In 1952, the station was located at three separate bases: in Alupka, Simeiz (Koshka Mountain) and Alushta. The main experimental base was located at Koshka Mountain, next to the Crimean Astrophysical Observatory. The radio telescope operating at 1.5 m wavelength was located there; during observations of the Sun in winter months, it could be used as a sea interferometer during the rising and setting of the Sun. The 7.5-m and 4-m radio telescopes were also located there, although the 4-m telescope was moved to Alushta in 1953.

Later, the B-1 radio telescope, operating at a wavelength of 3.5 m and based on an adaption of a radar station, was also constructed. Two such antennas formed an interferometer with a baseline of 160 m oriented East–West. Radiometers operating at 1, 2 and 2.6 m were mounted on the Ch-1 antenna (a parabolic dish without a means of steering). In addition, the facilities on Koshka Mountain included a small mechanical workshop, diesel and storage buildings, a cafeteria and an administrative building with accommodation, to which the heads of the station were moved from Alupka in 1953 and which also housed the construction group starting from that same year.

Two 3-m antennas operating as an interferometer at a wavelength of 50 cm were located at Alushta (but were moved to Koshka Mountain together with their radiometer in 1953), as well as a 4-m radio telescope operating at 23 cm wavelength.

First Investigations of the Supercorona of the Sun At the beginning of 1952, Vitkevich organised the Ashkhabad station for observations of the solar eclipse of

February 25, 1952. Chikhachev, G. G. Stolpovskii (assistant head of the station), the engineers T. M. Egorova, Rykhkov, Kalachev, and Medvedeva and others took part in this work. The observations were carried out using the Ch-1 antenna with four dipole systems placed at its focus, adjusted to receive at wavelengths of 1.1, 2.6 and 5.2 m. An original turning device supported by two tripods enabled tracking of the Sun during the eclipse.

The observations of this eclipse allowed Vitkevich and Chikhachev to obtain data on the distribution of the solar radio emission at metre wavelengths, and to determine the effective parameters of this emission together with their wavelength dependences.

After the successful completion of the eclipse observations, this antenna and its radiometer were moved to the Crimea (the village of Karabakh), where it was subsequently used for a series of observations carried out under the supervision of Chikhachev aimed at studying the radio emission of the active Sun.

Two new members of the Oscillation Laboratory joined the station in 1952— R. L. Sorochenko and V. A. Udal'tsov. Sorochenko modernised the 7.5-m antenna to enable observations of the Moon as well as the Sun at 10 cm. These measurements provided data about the daytime and nighttime refraction in the atmosphere, and also about the radio emission of the Moon itself.

As early as 1951, Vitkevich proposed the idea of investigating the solar corona by analysing radiation of the Crab Nebula source that had passed through the coronal material, when this source was located at various angular distances from the centre of the Sun.

Each year, the Crab Nebula approaches to within about 4.5° of the Sun on June 14–15. It is possible to estimate the parameters of the medium making up the corona by analysing the influence of the solar corona on the radiation passing through it from a background source. This could take the form of refraction, absorption or scattering of the radio waves due to inhomogeneities in the electron density in the corona. All these effects (except for the effects of small-scale structure) should be manifest more strongly with increasing wavelength. This made observations at metre wavelengths most suitable for such studies.

However, the Sun is a brighter source of radio emission than the Crab Nebula. The flux density of the radio emission of the quiescent Sun at a wavelength of 3 m exceeds 2×10^4 Jy[5], while the flux density of the Crab Nebula is about 2×10^3 Jy. Therefore, it would not be possible to distinguish the Crab Nebula's radiation from the stronger solar radiation as the nebula approached the Sun. An elegant method was proposed to attenuate the signal from the quiescent Sun, using the difference in the angular sizes of the Sun and the Crab Nebula source (30 and 6', respectively). The interferometer baseline was chosen so that the angular width of the interferometer's resolution was smaller than the angular size of the Sun, but larger than the angular size of the Crab Nebula source.

Vitkevich conducted the first such measurements in 1951 at 4 m wavelength using a sea interferometer with a baseline of 420 m (the interferometer resolution was 16').

---

[5]The unit of flux density used in radio astronomy is the Jy, 1 Jy = 10^{-26} W/(m² Hz).

The Crab Nebula was observed as it was rising. However, the Sun was active in the period of approach (from June 9 to June 22): local sources of radio emission strongly distorted the interference pattern due to the Crab Nebula.

A second set of observations was undertaken in 1952. An expedition under the supervision of Udal'tsov was organised to Ai-Petri, where it was possible to lay out long interferometer baselines. Two East–West interferometers operating at 6 m and 12 m wavelength and with baselines of 700 and 1200 m were set up. The 6-m observations were successful (the amplitude of the interferometer signal decreased), but it was not possible to carry out the 12-m observations due to a high level of interference. When analysing the results of these 1952 observations at Ai-Petri, Vitkevich concluded that the solar corona influenced the passage of radiation from the background source at distances of more than 4.5 solar radii from the centre of the Sun.

These observations of the Crab Nebula through the solar corona showed the effectiveness of this method. However, since it rained during the observations, the parameters of the radio telescope could have changed during the session. This raised the necessity of improving the observational methods. It was decided to conduct simultaneous observations at a single wavelength but with different interferometer baselines. In the case of scattering, the apparent angular size of a source should increase as it approaches the Sun, while this size should remain constant in the case of absorption. Therefore, in the case of absorption, as the source approaches the Sun, the variations in the interference pattern should be correlated on different interferometer baselines; in the case of scattering, the variations should be visible earlier on long baselines than on short ones. To study the form of the scattering region, it would also be desirable to have baselines at the same wavelengths but with different orientations.

At this time, Vitkevich had already introduced the term *supercorona* in his publications, referring to the outer region of the corona that was discovered due to its effect on other background radio sources.

Further Development of the Work   It followed from the first observations of the supercorona that further studies of the solar corona based on analysing signals passing through the corona required the construction of a large number of interferometers operating at various wavelengths and with baselines that differed in both length and orientation.

In 1951, none of the bases of the Crimean Station had areas that were suitable for such new interferometers. Therefore, a new experimental base was founded in 1953 near Katsiveli at the initiative of Vitkevich, and subsequently became the main base of the station. It is located to the west of Simeiz beyond Goluboi Bay, near the Black Sea Division of the Moscow Hydrophysical Institute of the Academy of Sciences.

In the period from 1952 to 1955, the number of workers at the station steadily increased from 20 to 90. Yu. A. Alekseev, A. N. Sukhanovskii, Yu. L. Kokurin, A. D. Kuz'min, M. A. Ovsyankin and Z. I. Kamenova began their work at the station in 1953–1955, and L. I. Matveenko in 1956.

To better organise the work being done at the station in this period, several scientific groups were founded, which carried out various observations and investigations

Fig. 1.8 Reflecting radio telescope used to conduct spectral observations at 21 cm starting in 1954

under the overall supervision of Vitkevich. The group of Udal'tsov performed experimental studies of the solar corona based on analyses of signals that had passed through the corona, and independently developed radio astronomy studies of the Crab Nebula; the group of Kokurin (to an appreciable extent independently) carried out experimental investigations of the ionosphere. The group of Chikhachev built instruments and carried out spectral studies of hydrogen lines. The construction of a spectral radiometer in this group was supervised by Sorochenko (Fig. 1.8). The group of Alekseev worked on experimental studies of the Sun at 1.5 m wavelength, including regular observations of the Sun during the International Geophysical Year. The group of Kamenova and Ovsyankin performed experimental investigations at wavelengths of 50 cm and 1 m of the sporadic radio emission of the Sun at frequencies of 80–120 MHz. A number of researchers also worked under the direct supervision of Vitkevich. In addition to the scientific and administrative groups, the Crimean station included a construction group (M. M. Tyaptin and others) and a group of mechanics.

Development of Methods and the Experimental Base In 1952–1956, a number of new methodological and constructional ideas were put forward by researchers at the station. Vitkevich proposed and developed a method for realising a wide-band radio interferometer, which was essentially a new means of creating an antenna system with a single beam providing high spatial resolution; a multi-channel radio interferometry system with two outputs, intended for determining the coordinates of a rapidly moving source whose intensity was changing; and an interferometer with a double modulation system (Fig. 1.9). Vitkevich and Udal'tsov also proposed a radio interferometer with a scanning antenna beam, which was realised by Udal'tsov in 1955–1956. Vitkevich and Sorochenko put forth the idea of creating a multi-element radio interferometer in the form of an equivalent diffraction grating, versions of

Fig. 1.9 Interferometer formed of two radio telescopes. The diameters of these parabolic dishes are 4 m. Observations at decimetre and centimetre wavelengths were carried out on this interferometer

which were put into operation by Udal'tsov in 1953 and by Matveenko in 1956. Udal'tsov developed a correlational method for polarisation measurements in 1957, bringing it into operation in 1960. Kuz'min proposed a method for changing the direction of an interference beam by changing the frequency of the heterodyne used. In 1954–1955, he developed noise generators and introduced them into use in measurements of the radio emission of the Sun and cosmic sources. In 1957, Matveenko devised a polarisation device in which the analysis of the received signals was carried out at an intermediate frequency, and brought into operation an interference triangle operating at 5 m wavelength in 1959.

A large antenna complex was constructed in the Crimea from 1952–1960, for radio astronomy observations under the supervision of Vitkevich. Old instruments were modernised as well, and a number of interferometers and receiving devices constructed.

Let us consider in more detail the construction at the initiative of Udal'tsov of a precise, transit, parabolic radio telescope with a diameter of 31 m (the RT-31), based on the earlier "eastern" RT-30 radio telescope. A special feature of this instrument (Fig. 1.10), which was unique for its time, was that it provided a large effective area suitable for use at wavelengths right down to 3 cm in a relatively simple way. The precise parabolic surface was obtained using a knife template, which was used to monitor the concrete and metal surfaces. The reflecting surface was sprayed with zinc. A platform supporting a movable cassette with feeds and receivers was installed on a vertical mast, above the focus of the paraboloid. This cassette could be moved either by hand or automatically, according to a specified programme, in a plane parallel to the aperture, within 4° of the axis in two orthogonal directions.

In 1958, Udal'tsov and Sukhanovskii devised an original system for the automatisation of radio astronomy observations, making it possible to carry out the entire

Fig. 1.10 Large fixed centimetre-wavelength radio telescope, with a diameter of 31 m

observing process in accordance with a specified programme (including the calibration of the radio telescope) without the participation of an operator.

Continuation of Studies of the Solar Supercorona Multi-faceted studies of the solar supercorona were carried out in 1953–1955 at metre wavelengths using the Crab Nebula as a source of radio radiation passing through the coronal material and the methods and resources described above. These investigations established the presence of an inhomogeneous scattering medium at large distances from the Sun, which increases the apparent angular size of the Crab Nebula source as it approaches the line of sight toward the Sun. The supercorona extends to several tens of solar radii, nearly to the orbit of Mercury.

Subsequent observations in 1956–1958 elucidated the role of the 11-year solar activity cycle in variations in the size of the supercorona. It turned out that the size of the supercorona increases as the solar activity grows. This correlation was revealed for both the inner and outer supercorona. The electron densities in inhomogeneities nearly doubles during periods of maximum solar activity. These observations showed that electrons concentrated in inhomogeneities in the solar supercorona are always present, in years of both minimum and maximum solar activity.

All the available data were used to construct models for the times of minimum and maximum solar activity. In the course of this work, asymmetry of the supercorona was discovered: the sizes of the inner and outer supercoronae are 1.5 and 2.5 times larger in the equatorial plane than they are perpendicular to this plane.

In 1956 and the following years, cases of the refraction of radio waves on large-scale inhomogeneities near the Sun were discovered, and used to estimate the parameters of these inhomogeneities. Focusing by these inhomogeneities was invoked to explain the growth in the intensity of a background radio source when its radiation passed through the corona, first observed in 1953. Matveenko detected flows of material moving with speeds in excess of 1000 km/s in 1960.

This series of studies was completed in 1962 with the construction of a two-component model for the solar supercorona, based in part on optical data on the mean electron density in the corona. In 1962, Vitkevich was awarded a diploma commemorating the discovery of the supercorona (granted on October 6, 1961, with the discovery date given as November 11, 1954).

Vitkevich and B. N. Panovkin (1931–1983) detected anisotropy in the scattering by analysing measurements of the scattering angle for the Crab Nebula at different orientations of the line connecting the source and the Sun relative to the interferometer baselines used, for observations at wavelengths of 3.5 m (1957), 5.8 m (1958) and 3.5 m (1958). A. G. Sukhova and M. A. Ovsyankin also measured the scattering angle for the Crab Nebula in various directions. Observations obtained in the following years both in the Soviet Union and abroad confirmed these results, which led to the discovery by Vitkevich and Panovkin of radial magnetic fields in the supercorona (the discovery diploma was issued on July 14, 1970, with the discovery date given as June 1957).

Studies of the Crab Nebula In these studies of the solar supercorona, the Crab Nebula was essentially used as a source of background radio emission. Therefore, nearly all the fixed radio telescopes of the Crimean Station were pointed at the Crab Nebula at some time, making it possible also to carry out specialised investigations of the nature of the radio emission of this interesting astronomical object.

The importance of polarisation observations came to the fore in the first half of the 1950s. The new hypothesis that the emission of various astronomical objects was synchrotron radiation, put forth and developed by the Soviet scientists V. L. Ginzburg, I. S. Shklovskii and G. G. Germantsev, as well as by many scientists abroad, shattered the then still predominant view from optical astronomy that essentially all electromagnetic radiation was thermal in nature.

A determining test in verifying whether observed emission is synchrotron radiation is the detection of linear polarisation in the optical and radio (see Essay 3). The most suitable object for such studies was the Crab Nebula, classified by R. Minkowski as a supernova remnant. This object had also been accessible to observations with medium-size antennas at the Crimean Station.

The first polarisation study of the Crab was undertaken in 1954, in both the optical (V. A. Dombrovskii and M. A. Vashakidze at the Byurakan Astronomical Observatory) and the radio, at wavelengths of 3.5, 5.8 and 7.5 m (V. A. Udal'tsov). A high degree of linear polarisation was detected in the optical ($\sim 15\%$), but no polarisation was detected in the radio, with an upper limit of 3%.

One possible reason for the absence of polarisation at metre wavelengths could be depolarisation of the signal due to Faraday rotation in the interstellar medium, as well as in the medium of the nebula itself. Testing this hypothesis required linear-polarisation measurements at shorter radio wavelengths, where Faraday rotation is manifest more weakly. It was desirable to carry out these measurements with a large radio telescope, in order to measure the degree of polarisation with an accuracy of a fraction of a percent. It was with these goals in mind that the 31-m parabolic telescope, suitable for observations at wavelengths down to 3 cm, was constructed (the RT-31).

A radiometer operating at 9.6 cm was used in the polarisation measurements carried out with the RT-31. In 1957, Kuz'min and Udal'tsov detected linear polarisation of the radiation from the Crab Nebula, equal to $(3 \pm 0.5)\%$. Polarisation was independently detected in that same year by radio astronomers in the USA at 3.15 cm. The detection of linear polarisation in both the optical and the radio represented convincing proof of the presence in the Universe of sources of non-thermal (synchrotron) radiation, demonstrated the non-thermal nature of the radiation of supernova remnants and illustrated that the radiation of the Crab Nebula had the same nature across its entire spectrum.

Udal'tsov carried out polarisation measurements of the Crab Nebula at 21 cm in 1959, using the RT-31 radio telescope and a specially constructed correlation polarimeter. These accurate measurements revealed linear polarisation at this wavelength as well, equal to $(11 \pm 0.5)\%$. It turned out that the polarisation position angles at 21, 9.6 and 3.15 cm displayed a linear dependence on the square of the wavelength. Thus, the effect of Faraday rotation was first detected, and used to measure the rotation measure in the interstellar medium and to demonstrate the depolarisation of the radiation of the Crab Nebula. This result also provided direct experimental evidence of the presence of regular interstellar magnetic fields.

A unique phenomenon occurred in 1955—an occultation of the Crab Nebula by the Moon. A special detachment was sent to the Pakhra Station of FIAN (see below), where radio astronomy studies at millimetre wavelengths were carried out near Krasnaya Pakhra (now the city of Troitsk), in order to observe this event. The complete phase of the occultation was observed on November 30, 1955 and January 24, 1956. Measurements were carried out using interferometers operating at 6 and 3.5 cm. As a result of these studies, Vitkevich and Udal'tsov detected an increase in the brightness toward the centre of the nebula and non-radial symmetry of the emission region—similar to that observed in the optical.[6]

Studies of the Radio Emission of the Sun Studies of the distribution of the radio brightness of the Sun in polar and equatorial regions carried out by Ovsyankin and Panovkin in 1955–1958 at wavelengths of 50 cm and 1 m showed that the shape of the Sun at 50 cm resembles an ellipse whose minor axis joins the two polar regions. In contrast to the situation at 1 m, an enhancement in the brightness was observed at this wavelength at a distance of 0.65 solar radii from the centre of the Sun, which is more weakly expressed in the polar regions. Vitkevich and Panovkin had discovered the "radio limb" of the Sun at 50 cm.

---

[6]The conclusion drawn previously that the Crab had different sizes in the optical and in the radio was erroneous, and due to insufficient accuracy in the measurements used. In subsequent observations, using the large telescopes of FIAN (the 22-m and DKR-1000 telescopes in Pushchino) and the antennas of the Deep Space Communications Centre, Matveenko, Sorochenko, V. S. Artyukh and others (1964 and 1974) demonstrated that the sizes of the Crab Nebula in the optical and radio were the same.

Based on an analysis of all the new solar radio data obtained by Ovsyankin and his collaborators, Panovkin proposed a new model with a temperature gradient in the inner corona of the Sun. He showed that it was not possible to explain the shift of the brightness maximum inside the optical disk and the growth of this shift with increasing wavelength using a simple, isotropic model for the Sun.

In 1957, Vitkevich, Kuz'min, Salomonovich and Udal'tsov made the first radio survey of the Sun at centimetre wavelengths, using the RT-31 telescope and scanning in declination with the antenna beam, together with radiometers operating at 3.2 and 9.6 cm. This revealed the presence of regions of enhanced radio brightness. A correlation was established between these regions and manifestations of solar activity in the optical (coronal condensations). Investigations continued by Vitkevich and Matveenko at wavelengths of 1.5 and 3 cm with polarisation sensitivity showed that some active regions were circularly polarised.

Vitkevich and M. I. Sigal established the presence of a correlation between the slowly varying component of the solar radio emission at 50 cm and coronal condensations in 1956. Vitkevich and his coworkers obtained new results in 1955–1960, including the discovery of a new type of solar radio flare, quasi-monochromatic phenomena on the Sun and small "peaks" of radio emission of various classes.

In 1959, after generalising the data on the total radio emission of the Sun obtained at 1.45 m during the International Geophysical Year (from August 1957 through August 1958), Alekseev came to an important conclusion—the sporadic radio emission of the Sun was due to local, primarily polarised, sources on the solar disk, with sizes, as a rule, less than $9'$.

Vitkevich and Alekseev obtained various important results based on polarisation measurements of the active Sun at 1.45 m, including results concerning the polarisation of the "peaks" mentioned above.

Development of Radio Astronomy Spectroscopy In 1951, American, Australian and Dutch radio astronomers detected nearly simultaneously the radio line of neutral hydrogen at 21 cm, which had been predicted earlier by the theoreticians van de Hulst and Shklovskii (see Essay 3). At the end of 1952, Chikhachev carried out a series of monochromatic investigations, aimed at developing a radio spectrometer operating near 21 cm, for studies of this hydrogen line. During 1953–1954, this radio spectrometer was developed by Sorochenko in Moscow, and brought to Katsiveli in Autumn 1954, where it was placed in the laboratory near the "western" V-3 radio telescope.

The first recording of the 21-cm hydrogen line at the Crimean Station was obtained on December 28, 1955; these were the first successful radio astronomy spectroscopic observations carried out in the Soviet Union. More than three years went by before regular observations were established, however. During this period, the specific requirements of spectral radio astronomy observations were studied, instruments prepared and the radio telescope outfitted; in particular, an automated tracking system incorporating electro-mechanical coordinate translation, originally developed for the 22-m radio telescope in Pushchino (see below), was installed.

In 1954, Sorochenko, Vitkevich and Chikhachev attempted to detect line radio emission from the CH radical, which, according to the calculations of Shklovskii (which subsequently were shown to be too optimistic), should have been emitted at a frequency of 3180 MHz. For this purpose, they used their accumulated experience to develop a spectrometer operating at 10 cm wavelength, which was then installed on the 7.5 m radio telescope. The results of spectral observations toward the centre of the Galaxy were negative: the sought-for CH line was not detected in the frequency range (3190 ± 12) MHz (with the antenna temperature of the line being <30 K).

Sorochenko began a large series of studies in a complex region of the Galaxy— Cygnus X, including measurements of the distribution of neutral hydrogen. It turned out that Cygnus X is located inside the nearest spiral arm, and consists of two main condensations of ionised hydrogen located at distances of 1.5 and 3 kpc. Analysis of both radio and optical studies were used to find the distances to four regions of ionised hydrogen near the Galactic plane, between Galactic longitudes of 75 and 89°. It was shown that an observed cluster of optical emission nebulae in Cygnus X coincided with a region of enhanced radio emission, and was due to the projection of several HII regions located at various distances.

This group of spectral studies at the Crimean Station formed a centre and a school for spectral radio astronomy in the USSR. Many specialists from other institutions came to carry out spectral investigations and to obtain experience in instrumental development, including the now widely known radio astronomers N. S. Kardashev (Space Research Institute), T. A. Lozinskaya and M. I. Pashchenko (Sternberg Astronomical Institute), T. M. Egorova (Main Astronomical Observatory) and G. T. Tovmasian (Byurakan Astronomical Observatory). Sorochenko became the scientific supervisor of spectral studies in the Radio Astronomy Laboratory in 1960 (after the departure of Chikhachev).

Ionospheric Studies Using Radio Astronomy Methods Observations of the inhomogeneous structure of the ionosphere based on the analysis of radio emission from the most intense cosmic radio sources (Cassiopeia A, Cygnus A, Virgo A and the Crab Nebula) after its passage through the ionosphere begun earlier were continued starting in 1951. Beginning in 1954, these studies were carried out by the group of Yu. L. Kokurin.

A series of studies of small-scale inhomogeneities in the ionosphere were conducted as part of the programmes of the International Geophysical Year and the International Year of the Quiet Sun. The dimensions, shapes and motions of inhomogeneities with characteristic sizes of about several kilometres were studied in detail, as well as their daily and seasonal behaviour and their connection with the state of the Sun. This work was carried out using observations of Cassiopeia A, Cygnus A and Virgo A made with spatially separated receivers operating at a wavelength of 6 m.

Small-scale ionospheric inhomogeneities were also studied using sounding measurements made with a triangle formed of three rhombic antennas operating at 7– 8 MHz. This yielded the interference patterns of magneto-ionic components of the radio signal reflected from the ionosphere. On this basis, it was proposed that fading of the signal was observed, and the fluctuations in the electron density due to small-scale ionospheric inhomogeneities were estimated (from a joint analysis of the results of transmission and ionosphere-sounding measurements).

Another large series of studies continuing until 1969 was devoted to studies of the large-scale structure of the ionosphere. This work was a direct continuation of investigations begun in 1949. On their basis, Vitkevich and Kokurin showed that the regular component of the vertical refraction in the ionosphere is due to inhomogeneities with sizes of about 200 km in the F layer. The speeds of the motions of these inhomogeneities in the ionosphere were measured (50–100 km/s). Vitkevich estimated the ratio of the total number of electrons contained in inhomogeneities to the total number in a column with unit cross section for the unperturbed ionosphere with inhomogeneities with sizes of 5 km, based on the results of interference observations obtained in 1951–1956.

Based on calculations of the dependences of the vertical and horizontal refractions on the zenith angle at 4 m wavelength carried out in 1958, Kokurin identified an appreciable difference between these dependences for two models for the non-uniform structure of the ionosphere (wavy and cloudy). He pointed out that this difference could serve as a basis for discriminating between ionospheric models.

A series of simultaneous refraction measurements at zenith angles of 0–30 and 77–90° carried out in 1955–1959 enabled Kokurin, Sukhanovskii and Alekseev to obtain new data on large-scale ionospheric inhomogeneities, in particular on their scales, structure and behaviour throughout the day. Ovsyankin also actively took part in all the ionospheric studies carried out after 1957.

Radio interference methods have found applications in a wide range of radio astronomy and applied problems.

Observations of the First Soviet Space Rockets In 1958, the Crimean Station of FIAN faced an important and extremely complex task—determining the flight trajectory of a rocket heading towards the Moon. The construction of the measurement complex was supposed to be completed over several months. The head of the Oscillation Laboratory of the FIAN, A. M. Prokhorov, defined the task at hand. It was necessary to construct a high-sensitivity, precision interferometer operating at a wavelength of 1.76 m. The work was monitored directly by M. V. Keldysh and S. P. Korolev, who arrived and acquainted themselves with the work at the site. An operative headquarters was established, headed by Vitkevich.

The antennas of the radio spectrometer, which seemed huge and, in any case, immovable, were shifted over several hours and mounted with an accuracy of several millimetres. New feeds (Yu. P. Ilyasov) and low-noise amplifiers (Matveenko) were developed and constructed, the cable system was laid and the necessary equipment assembled. The radio interferometer was ready in Autumn; it was aligned using radio astronomy methods—observations of cosmic radio sources.

On January 2, September 12 and October 4, 1959, this radio interferometer was used to determine the flight trajectory of the rocket with very high accuracy for that time—about a minute of arc (absolute error). The time when *Luna-2* would land on the Moon was determined, and the automated radio measurement complex was used to determine the landing site, near the crater Archimedes. The results of this unique experiment were published (Vitkevich, Kuz'min, Matveenko, Sorochenko and Udal'tsov).

Together with this work of the Crimean Station of FIAN, radio astronomy investigations were carried out on the eight-antenna radio telescope of the Deep Space Communications Centre (see Essay 9). The preparation of the equipment complex and organisation of the observations was supervised by Matveenko. In 1960–1961, equipment operating at 32 and 8 cm was developed with the direct participation of G. Ya. Gus'kov. The first observations of cosmic radio sources at 32 cm were carried out in the Summer of 1962, and the first observations at 8 cm in the Summer of 1963. These observations were the first to use a quantum paramagnetic amplifier (Matveenko, V. B. Shteinshleiger).

At the stage of bringing the equipment into operation, a number of PhD students at the Sternberg Astronomical Institute joined the effort (see Essay 3). The ability to join the forces of these two organisations with the friendly participation and help of V. A. Kotel'nikov and his group at the Institute of Radio Physics and Electronics made it possible to accelerate the work. Detailed measurements of the radio telescope's beam were conducted, and the presence of absorption by thermal electrons in the spectrum of the Seyfert galaxy NGC 1275 was established (Matveenko).

In November 1958, the Crimean Station of FIAN was reorganised into the Crimean Scientific Station, and work in radio astronomy began to be taken over by the newly constructed Radio Astronomy Station in Pushchino. In 1962, the Crimean Station became part of the laboratory of N. G. Basov, and the topic of the work there changed; the leading radio astronomers who had worked there moved to Pushchino.

1.3 The Radio Astronomy Station in Pushchino. Founding and Development

The Moscow Division of the Radio Astronomy Section As early as November 1951, at the initiative of Khaikin in the framework of a programme to organise a solar service in the Soviet Union, the USSR Academy of Sciences and two ministries were given the task of preparing proposals for the planning and construction of large reflecting telescopes for radio observations of the Sun and other cosmic radio sources at centimetre and millimetre wavelengths. For the first time, the construction of a radio telescope with the participation of industry was discussed. The possibility arose of forming an experimental basis for developing radio astronomy at millimetre wavelengths. Khaikin offerred the development of this direction to Salomonovich, and the development of a radio telescope operating at millimetre wavelengths to Kalachev. In 1952, the group of Salomonovich included the radio technicians E. K. Karlova and N. A. Amenitskii and the mechanic S. K. Palamarchuk, with U. V. Khangil'din joining slightly later.

In 1952–1954, a radiometer designed to operate at 8 mm with a balance mixer based on silicon diodes and an electrodynamic modulator was devised. At the same time, Kalachev developed a turning device with an equatorial mount for a glass parabolic surface (2-m diameter) taken from a searchlight. This radio telescope, the RT-2, was apparently the first Soviet radio telescope to have an equatorial mount.

Adjustments in declination were carried out by hand using a theodolite. The radio telescope was manufactured entirely at FIAN by workers in the corresponding shops in the Radio Astronomy Section (A. A. Levin, I. G. Evtyukhov, P. D. Lebedev and A. V. Rybakov).

By the Summer of 1954, the RT-2 radio telescope together with a cooled radiometer was ready to be sent to Novomoskovsk (in the Dnepropetrovsk region), where it was proposed to observe a solar eclipse on June 30 (a large group of researchers from the Crimean Station also took part in these observations, using radio telescopes operating at metre and decimetre wavelengths). Several days before it was to be shipped, the two-metre surface cracked as it was being mounted into the frame of the radio telescope. The situation was saved by Ya. N. Fel'd, who kindly loaned a similar but slightly smaller parabolic surface. Yu. N. Pariiskii actively took part in the adjustment of the radio telescope and the observations that were conducted during the eclipse. These first observations of a solar eclipse at millimetre wavelengths in the USSR revealed an enhancement in the radio brightness of the solar limb.

In 1953, two Masters' students of Salomonovich, and then two PhD students of Khaikin, N. V. Karlov and T. A. Shmaonov, began their work in the Moscow Division of the Radio Astronomy Section. At that time, in connection with the move toward observing at shorter wavelengths and the need to observe relatively weak sources, the question of the limiting sensitivities of radio astronomy equipment became quite critical. This question had been examined from a theoretical point of view by F. V. Bunkin (a PhD student of S. M. Rytov) and by Karlov; the results of their studies subsequently became classic.

With the aim of selecting an optimal modulation frequency, Salomonovich and Smaonov experimentally studied the spectrum of the low-frequency noise of a radiometer. For the first time, a non-mechanical means of modulation was applied at centimetre wavelengths, based on the Faraday effect in a ferrite rod, and the first model of an automated zero radiometer was developed (Salomonovich and Karlov).

In 1955, the RT-2 radio telescope operating at a wavelength of 8 mm was installed at the site of the Solar Radio Laboratory in Krasnaya Pakhra (Institute for Terrestrial Magnetism, the Ionosphere, and Radio-wave Propagation of the Academy of Sciences). In 1955–1958, Salomonovich carried out a series of solar observations, including observations of the partial eclipse of 1956. These observations made it possible to determine the brightness temperature of the Sun and the dependence of the slowly varying component on the activity index. The phase dependence of the brightness temperature of the lunar radio emission was detected for the first time. In connection with this last result, the presence of a second harmonic of the periodic dependence of the varying component on the phase was detected, and was consistent with the theory worked out at that time by V. S. Troitskii. Radio astronomy methods were also used to study the absorption of 8 mm radiation in the atmosphere as a function of the meteorological conditions (O. M. Ataev, Salomonovich).

In the beginning of 1953, a new group focusing on studies of monochromatic radio emission was formed in Moscow under the leadership of Chikhachev. This group included Sorochenko, Yu. L. Sverdlov, S. A. Zaitsev and V. I. Kruchinenko.

In a comparatively short time, they had built a complex and, in many ways, quite unusual spectrometer designed for studies of 21 cm radio emission.

Kuz'min joined the Radio Astronomy Section in 1954; earlier, he had specialised in measurements of electrical noise signals and the construction of high-sensitivity receiver-amplifier schemes, being the author of the book *Measurements of the Noise Coefficients of Receiving and Amplifying Devices* (Moscow/Leningrad, Gosenergoizdat, 1955). A group working on the development of methods for measuring the intensities of cosmic radio sources and the construction of radiometers with low noise levels was formed under his supervision (L. A. Levchenko, M. T. Levchenko, V. S. Borodacheva and A. N. Khvoshchev). The first noise generators in the Soviet Union were built. In 1957, the group finished the development and construction of a high-sensitivity polarisation-sensitive radiometer designed to operate at 9.6 cm, which had the high fluctuation sensitivity for that time of about 0.15 K.

Salomonovich, Karlov and Amenitskii devised a new generation of radiometers operating at 3 cm, which were subsequently installed on the radio telescope in Pushchino, and then on the RT-31 radio telescope in the Crimea in 1957 (see above). This same group began the development of a polarisation-sensitive radiometer for observations at 8 mm, based on the use of ferrites, designed to measure all four Stokes parameters. This work, carried out by U. V. Khangil'din, led to the establishment of an installation for studies of solar active regions.

The first polarisation radiometer indicated above was tested during observations of the solar eclipse of April 24, 1958 on a radio telescope equipped with a parabolic surface with a diameter of 90 cm and an equatorial mount located on Hainan Island (People's Republic of China), where collaborative observations were carried out by the Academy of Sciences of the USSR and scientists of the PRC. Salomonovich, Khangil'din and Amenitskii participated in these observations.

N. A. Lotova joined the Radio Astronomy Section in 1955, working on the theory of the scattering of radio waves in the interplanetary medium under the supervision of Vitkevich. In connection with this, the work of the construction group of Kalachev was very productive.

Thus, the Moscow Division of the Radio Astronomy Section, and further the Radio Astronomy Laboratory, played a very important role in the development of radio astronomy techniques and equipment and the construction of radio telescopes.

Construction of the RT-22 Telescope. First Observations in Pushchino The large amount of experience with radio astronomy investigations obtained in the 1950s led to the determination of the main requirements for radio telescopes. The question of expanding into the range of millimetre and short centimetre wavelengths became pressing. Observations at precisely these wavelengths could more easily be used to compare the results of radio and optical studies. This required a radio telescope with a large effective area to enhance its sensitivity and angular resolution. The construction of such an instrument was not a simple task, however. There was no experience in constructing large, steerable, precision surfaces either in the Soviet Union or abroad.

None of the industrial organisations at that time was willing to take on the construction of such an instrument, citing lack of experience. Therefore, the development of a draft design for a steerable radio telescope with a parabolic surface and a diameter of 16 m was entrusted to FIAN in 1955, together with a number of other organisations, each of which would be responsible for the development of various individual systems of the radio telescope.

The work on the draft design of this 16 m radio telescope, in which Kaidanovskii participated, in addition to Salomonovich and Kalachev, was completed in 1952. Starting in February 1954, workers at FIAN and a number of organisations and ministry industries carried out the technical and operational design of an experimental model of the radio telescope. Based on the results of the draft design, after analysing the possibilities for constructing a supported, steerable device and the framework for the reflecting surface, it was decided to increase the diameter to 22 m and construct the radio telescope that is now known as the RT-22 of FIAN.

A division concerned with the development and exploitation of radio telescopes was formed in the Radio Astronomy Section of the Oscillation Laboratory of FIAN. The scientific supervisor of this division was the main construction engineer in the section—Kalachev. G. G. Stolpovskii was the assistant head of the section for some time. Attentive interest in this work was expressed by the assistant director of FIAN, M. G. Krivonosov, as well as the head of the Oscillation Laboratory, Prokhorov.

There were many difficulties in constructing the RT-22 radio telescope. On the scientific and technical side, it was necessary to solve the fundamentally new problem of constructing a reflecting surface for an antenna that was 22 m in size with an accuracy of a fraction of a millimetre. In addition, when the dish, which had a mass of more than 50 tons, was tilted, the deformations of the shape must likewise not exceed a fraction of a millimetre.

Kalachev and Salomonovich and their teams solved the problem of how to build a precise, large reflecting surface using a new and original type of construction— attaching an accurate reflecting surface to a rigid but crude frame using studs. The accuracy of the surface was achieved by adjusting the length of the studs, and its rigidity by an optimal choice of construction.

The development and construction of an instrument such as the RT-22 did not present a complex problem in those years only from a mechanical and surface-accuracy point of view. It was necessary to determine how to receive and channelise the radio signal on scales that had never been used before; high-sensitivity receiver/recorder equipment operational at a wavelength of 1 cm was required; the pointing and tracking system was supposed to steer an instrument with a mass of more than 700 tons with an angular accuracy of no worse than $10-20''$ with an altitude–azimuth mount, which required translation from equatorial coordinates. Specialists from a number of industrial organisations also took part in the development of the radio telescope (A. A. Pistol'kors, L. D. Bakhrakh, K. I. Mogil'nikova, M. I. Grigor'eva, I. V. Vavilova, G. Yu. Pogozhev, N. M. Yakimenko, Yu. N. Semenov, V. A. Vvedenskii, P. V. Dobychin, A. N. Kondrat'ev, N. Ya Bulkin and others).

When the designs for the supporting and turning structure arrived at the construction division of the manufacturing plant, the constructional engineers there pointed out the strong similarity of the support and turning system for rotation in azimuth with the design for the main gauge of a linear ship that had been manufactured earlier and never used, and suggested that this prepared structure be used, which saved time and resources. A team of constructional engineers from FIAN headed by Kalachev went to the plant, reworking over several weeks the adjacent nodes for the support and turning system for rotation in hour angle.

The construction of the RT-22 radio telescope played an important role in the beginnings of the Radio Astronomy Station of FIAN in Pushchino, which now carries out primarily radio astronomy studies.

In 1954, the question arose of finding a suitable location for the RT-22, together with a complex of scientific and technical buildings, living quarters and buildings housing the equipment necessary for its use.

It was first proposed to place the new radio telescope on a small area located on the territory of the Institute of Terrestrial Magnetism, the Ionosphere and Radiowave Propagation near Moscow. However, in 1955, Vitkevich and Chikhachev suggested that another new cross radio telescope operating at metre wavelengths be built. This radio telescope (see below for more details) needed an area 1 km by 1 km in size, and a much larger area was required for both radio telescopes. Therefore, FIAN made the decision at the beginning of 1955 to built a radio astronomy station where both radio telescopes would be located somewhere near Moscow.

Special groups under the leadership of Stolpovskii and M. M. Tyaptin were formed to search for a suitable site, and investigated a number of places in the southern part of the Moscow region, as well as in the Kaluga, Ryazan and Tula regions.

As a result, a site on the banks of the Oka River was chosen, where it was proposed in the future to house a scientific centre of the USSR Academy of Sciences (now the city of Pushchino).

The construction of the Radio Astronomy Station of FIAN helped prepare the production–construction base for the future scientific city, which also simplified the solution of a number of day-to-day and social questions concerning the lives and work of the workers at the station.

It is traditional to take the birthday of the Radio Astronomy Station to be April 11, 1956, when the permission to construct the RT-22 radio telescope was granted. Thus, FIAN became the first scientific institution to begin the creation of a scientific base at the site of the future scientific city of Pushchino.

The new location began to take shape under the general supervision of Vitkevich, who was at that time the head of the Radio Astronomy Section of the FIAN, and his assistants Salomonovich and Kovalevskii. The Oka Station was organised to monitor the construction and assembly of the equipment, and also the preparation of the first scientific studies using the cross and RT-22 radio telescopes (subsequently the continuously operating Oka Radio Astronomy Station of FIAN).

In February 1960, the Radio Astronomy Section of the Oscillation Laboratory was separated off to form the independent Radio Astronomy Laboratory of FIAN, consisting of the Oka and Crimea Radio Astronomy Stations, as well as the Moscow group. Vitkevich became the head of this laboratory, simultaneously being named the head of the Oka Radio Astronomy Station, with Kovalevskii as his assistant.

Fig. 1.11 Assembly of the reflecting surface of the FIAN 22-m radio telescope using a knife template at Pushchino (1957–1958)

While the development and manufacture of the two large radio telescopes was underway, Vitkevich continued his investigations of the solar supercorona in Pushchino.

Using several small antennas from obsolete radar stations, Vitkevich and Panovkin conducted observations that established that the solar supercorona has a radial filamentary structure. Since the motion of charged particles is "frozen" to the magnetic-field lines, they concluded that the Sun had a radial magnetic-field structure. This was the first discovery made in Pushchino, which was issued a discovery diploma in 1970, with the discovery date given as June 1957.

The foundation of the RT-22 was laid at the end of 1956. The construction of the laboratory buildings and living accommodation together with the complex of technical buildings was begun (G. V. Sovkov, A. B. Ulanovskii, Tyaptin).

The work was greatly expanded in 1957–1958. Teams under the supervision of V. I. Emelin and I. D. Krongauz assembled the metal elements and mechanisms arriving from the plants, and assembled and adjusted the control system (the groups of M. A. Vestfal' and V. A. Vvedenskii). The young specialist of the Radio Astronomy Station G. G. Basistov also participated in the adjustment of the control system.

Considerable help in the creation of the Radio Astronomy Station was given by the director of FIAN D. V. Skobel'tsyn, his assistants M. G. Krivonosov and I. F. Kalakin, the head of the Oscillation Laboratory and Corresponding Member of the Academy of Sciences A. M. Prokhorov and the secretary of the Party organisation of the laboratory N. G. Basov.

The construction of an accurate surface proved to be technically extremely complex. This work was brilliantly carried out in an original way by the team of V. N. Kondrashov under the supervision of Kalachev (Fig. 1.11). In November 1958, the surface achieved a record accuracy for that time—0.3 mm (rms error), which was provided by manufacturing the reflecting surface of the radio telescope using a precision template. This painstaking work, conducted under the dome of a specially constructed technical shop with a removable roof, occupied about half a year. On the morning of November 18, 1958, the multi-ton 22-m reflector was lifted from the assembly building using a travelling gantry crane, transported to the already constructed support–turning structure, placed onto this structure and fixed in place. The entire operation lasted two hours.

When the assembly of the RT-22 was completed and the telescope was making its first tests and observations, Kuz'min joined the group of Salomonovich, which had been entrusted with becoming acquainted with this new instrument, actively participating in the adjustment and measurement of the parameters of the RT-22.

By the end of 1958, the reflector axis was installed in the calculated position using a theodolite, the surface had been additionally adjusted, all the cable lines had been assembled and all the drive mechanisms aligned and adjusted. The feed supports were installed at the beginning of 1959, and the first version of the radiometer block, with receivers and feeds operating at 3 cm and 8 mm, mounted at the focus of the RT-22. Various methods for illuminating the main surface were considered during the designing of the RT-22. Salomonovich and N. S. Soboleva (1955) looked into various aspects of the electrodynamics of a two-mirror Cassegrain system. The draft design also considered a modern (operating at several wavelengths) waveguide–horn feed system. However, as a first version, a simple microwave block operational at wavelengths of 3.2 and 0.8 cm with waveguide feeds was developed and used (Fig. 1.12).

The radio telescope became operational at the beginning of May 1959. At the meeting of the A. S. Popov Scientific Society on May 7, 1959, which was dedicated to radio studies, the first recording of the solar radio emission at 8 mm was demonstrated. The recording of the passage of an arched active region across the solar disk was used to estimate the width of the antenna beam, which was 2′, close to the calculated value.

The experience accumulated during the construction of the FIAN RT-22 was later used to build and equip the analogous RT-22 radio telescope of the Crimean Astrophysical Observatory. In a resolution made on April 21, 1977, the Presidium of the USSR Academy of Sciences awarded Salomonovich and Kalachev (Fig. 1.13) the A. S. Popov Prize for outstanding achievements in the construction and study of millimetre-wave radio telescopes. This new instrument opened wide opportunities for radio astronomy investigations at centimetre and millimetre wavelengths.

In the first systematic observations on the RT-22, carried out in 1959–1960, the two-dimensional radio brightness distribution over the solar disk was measured at 3 cm and 8 mm. This provided the first information about the dynamics of radio-emitting active regions, and also made it possible to identify flares of the millimetre-wavelength radio emission (Salomonovich). Khangil'din detected a region of reduced 8 mm radio brightness on the solar disk associated with protuberances.

Soon after these first observations, measurements of the 8 mm radio brightness distribution across the disk of the Moon began, as well as systematic measure-

1 Radio Astronomy Studies at the Lebedev Physical Institute 33

Fig. 1.12 The 22-m radio telescope of FIAN. The first complex of radiometers (0.8 and 3.2 cm) is at the focus of the parabolic surface

Fig. 1.13 A. E. Salomonovich and P. D. Kalachev in the control room of the 22-m radio telescope (1959)

ments of the phase dependence of the lunar brightness temperature at various wavelengths. These measurements, in which Salomonovich, Amenitsii, Khangil'din, A. M. Karuchun, V. N. Koshchenko, R. I. Noskova and B. Ya. Losovskii participated, yielded results about the degree of constancy (in the case of the Sun) and the magnitude of the regular behaviour (in the case of the Moon) of the shifts of the centres of gravity of the radio emission over the disks of these objects.

After this first series of measurements, Salomonovich, Kuz'min and Karachun electrically adjusted the position of the RT-22 feeds and carried out the first studies of the telescope's parameters. The adjustment and measurement of the parameters of the RT-22 required the solution of a number of scientific and technical problems, which led to the development of new methods based on the use of cosmic radio sources. The considerable experience acquired in this work led Kuz'min and Salomonovich to write the monograph *Radio Astronomy Methods for the Measurement of Antenna Parameters* (Moscow: Soviet Radio), which was published in 1964 and translated into English in the USA in 1964. These methods were subsequently widely used in practise to measure the parameters of the large antennas used for radar and space communications.

An original method for measurements in the near zone—the defocussing method (Salomonovich, B. V. Braude, N. A. Esepkina)—was also used to adjust the telescope and estimate its effective area.

In January–April 1960, Kuz'min, Salomonovich and others carried out a survey of discrete radio sources at 9.6 cm on the RT-22 radio telescope, compiling the first radio catalogue in the Soviet Union, which contained 56 discrete sources, 34 of them observed for the first time at centimetre wavelengths. In June 1960, Karachun, Kuz'min and Salomonovich conducted observations of several radio sources at the shorter wavelength of 3.2 cm. It was discovered that the spectrum of the source Cygnus A becomes steeper at centimetre wavelengths. Based on this break in the spectrum, Kardashev, Kuz'min and S. I. Syrovatskii derived one of the first estimates of the age of this radio source—about 0.5 million years.

A group of Soviet radio astronomers participated in a Soviet–American symposium dedicated to problems in radio astronomy in 1962. Well known scientists from both the USSR and USA actively took part in this symposium (Fig. 1.14).

Construction of the DKR-1000 Radio Telescope In 1959, the construction of another large-scale instrument began at FIAN—the DKR-1000 metre-wavelength radio telescope. This was proposed by Vitkevich as a way to further develop studies of the solar supercorona and cosmological studies.

The discovery of the solar supercorona made detailed investigations of the structure and nature of circumsolar space of considerable interest. This required an increase in the number of radio sources that could be observed as they passed behind the supercorona, which motivated the need for a radio telescope with a large collecting area and high resolution.

At the same time, the problem arose of "counting the radio stars," as the cosmological problem of investigating the distribution of galaxies in the Universe was called at that time. The development of this line of research, which was extremely

Fig. 1.14 Participants of the Soviet–American radio astronomy symposium (USA, 1962). *Lower row (from left to right)*: G. G. Getmantsev, F. T. Haddock, two unidentified women, R. Minkowski, V. V. Vitkevich, O. Struve, R. L. Sorochenko, J. Fisher, G. Keller, A. D. Kuz'min, R. N. Bracewell, F. D. Drake; *middle row*: C. M. Wade, E. F. McClain, V. A. Sanamyan, P. D. Kalachev, G. Stanley, A. Barrett, H. Weaver, G. Swenson, C. Mayer, D. S. Heeschen, J. Kraus; *upper row*: G. Field, T. Menon, R. Seeger, L. Woltjer, A. Sandage, A. Lilly, A. Blaauw, F. Kahn, B. Burke

important for studies of the evolution of the Universe, required observations of a large number of sources. Meter wavelengths were most suitable for this purpose, since it was possible to achieve high sensitivity using comparatively simple technical methods, and high resolution could be provided by constructing the telescope in the form of an extended cross. It was also possible to match the sensitivity of the instrument for source counts with the telescope's resolution.

The mechanisms generating the gigantic energies of extragalactic sources already deeply interested scientists at that time. In this connection, as well, it was important to study the structures and spectra of as many radio sources as possible.

In 1956, Vitkevich aided by Chikhachev proposed to construct a Mills-cross-type radio telescope, but with reflectors in the form of parabolic cylinders, each 1000×40 m in size. Estimates indicated that such an instrument could detect several thousand discrete sources. The high angular resolution required for studies of the structure of radio sources could be obtained by adding comparatively small movable antennas to the main radio telescope, thus forming an interferometric system. This proposal was approved by the head of the Oscillation Laboratory of FIAN, Prokhorov, who also subsequently supported the construction of the metre-wavelength cross telescope.

Kalachev rapidly came up with a draft scheme for the project, according to which the east–west arm of the antenna would be rotated in hour angle using a synchronised electrical shaft containing 37 motors, one for each of the parabolic 40-metre trusses. It was planned to use an electrical phasing method to change the direction of reception for the north–south arm of the antenna to be consistent with that for the east–west arm.

During the construction of the radio telescope, Kalachev, V. P. Nazarov and Tyaptin spent an appreciable amount of time at the construction site, supervising the assembly process. The construction of such a gigantic structure was a complex technical task that was being tackled by the constructor engineers of the Radio Astronomy Laboratory for the first time: nobody in either the Soviet Union or abroad had previous practical experience with such a project.

The financing for the construction of the radio telescope was rather limited. The available resources were sufficient only to carry out the most important aspects of the work. The further development of the radio telescope system, in particular, of the receivers and control computer, had to be carried out later.

Essentially simultaneously with the mechanical construction, the electrical wiring for the east–west antenna was strung up, and the necessary electrical and radio frequency cables laid. At the suggestion of Alekseev and A. M. Veselov, the drive motors were used in an asynchronous regime, and equipped with protection against discrepancies between the positions of the trusses.

The development of the microwave system of the cross radio telescope began in 1957 with the construction at the Oka Station of a section of a feed with an operating wavelength of 3.5 m under the supervision of Kuz'min. This work was carried out by Ilyasov, aided by Kruchinenko and L. A. Levchenko.

During the analysis of the perspective use of the telescope, it became clear that narrow-band feeds would limit the observational capabilities of the future instrument; in addition, the use of such feeds would create complex problems associated with the movement or reassembly of the feeds in order to change the observing wavelength. Based on the broadband oscillators that had been designed by G. Z. Aizenberg and V. D. Kuznetsov, in 1957, Kuz'min and Ilyasov proposed the use of a broadband feed that could operate at 30–120 MHz.

A model of the V-3 section of the antenna was constructed in 1958, and was used to study the parameters of this section of the feed and work out the best means for its construction.

The broadband oscillators and a co-phased, parallel-storey design for the addition of the signals provided unique opportunities to carry out observations simultaneously within a range of two octaves. The instrument was accordingly named the *Diapozonnii krestoobraznii radioteleskop* ("Broadband Cross Radio Telescope", DKR-1000).

The design work for the radio telescope proceeded in parallel with the assembly and adjustment work. Much was done to organise this work by Kovalevskii, who was the assistant head of the station at that time, as well as by N. P. Vladimirov, N. F. Kotov and K. I. Stepnov.

The construction of the V-3 antenna reflector was carried out on schedule. The first of a series of feed sections consisting of eight oscillators was installed in the first span of the antenna, and the first observation of the passage of a discrete source was conducted on October 15, 1961.

The installation of the remaining feed sections all along the antenna began in Spring 1963. The alignment and adjustment of the electrical lengths of the trunk cables (500 m) to an accuracy of better than 1 cm—carried out by Ilyasov, aided by A. V. Afon'kin, V. T. Solodkov, A. N. Ivanov, N. N. Lapushinskii and others—was finished by the middle of 1964. By the same time, V. I. Vlasov, V. S. Medvedeva and T. I. Gavrilenko had manufactured a two-frequency radiometer that could operate at 30–120 MHz.

Fig. 1.15 DKR-1000 cross radio telescope of FIAN: North–South arm

Fig. 1.16 DKR-1000 cross radio telescope of FIAN: East–West arm

By the end of 1964, the V-3 section of the antenna was ready for scientific observations, which were initiated by V. S. Artyukh and R. D. Dagkesamanskii under the supervision of Vitkevich. The adjustment of the north–south antenna was finished and the antenna ready for observations at the end of 1965 (Figs. 1.15, 1.16).

Development of Studies Using the RT-22 in the 1960s and 1970s The RT-22 was actively used for research on the Sun, Moon, planets and other cosmic radio sources at millimetre and centimetre wavelengths.

Losovskii obtained images of the Sun at 2 cm before and after the solar eclipse of February 15, 1961.

As is noted above, the construction of a radio telescope with high resolution enabled investigations of the two-dimensional radio brightness distribution over the lunar disk at 8 mm, as well as at some centimetre wavelengths. These studies reliably established the presence of a latitude distribution of the brightness temperature, indicating a decrease in the surface temperature toward the poles. Lunar radio maps were used to estimate the mean dielectric constant ($\epsilon \simeq 2$). Estimates provided evidence that the material making up the lunar surface had a low density and thermal conductivity, and also a small loss angle (Salomonovich, Koshchenko, Losovskii).

In 1961, A. G. Kislyakov, Losovskii and Solomonovich detected a modest enhancement of the brightness temperatures at 4 and 8 mm in the region of the lunar "seas" compared to the region of the "continents." In 1963–1965, Losovskii and Salomonovich conducted a prolonged series of measurements at 0.8 and 1.35 cm. The results showed that the surfaces of the lunar seas and continents have slightly different physical characteristics, and Losovskii later concluded that, indeed, they have different chemical and mineral compositions. These conclusions reached on the basis of radio astronomy measurements were later confirmed in direct studies of the Moon using spacecraft.

By the completion of the construction of the RT-22 in 1958, a group of American radio astronomers had published a paper reporting that the brightness temperature of Venus was unexpectedly high—around 600 K. This contradicted the generally accepted view at that time that the planet Venus had physical conditions that were roughly the same as on the Earth. Kuz'min and Salomonovich developed a broad programme of studies of Venus and other planets, which was also very timely in connection with the flights of various spacecraft to Venus. FIAN proposed to study the possible usefulness of carrying out parallel terrestrial measurements of the radio emission of Venus at all accessible wavelengths, and this was viewed favourably by the President of the USSR Academy of Sciences, Academician M. V. Keldysh. Thanks to the support of the chairman of the Scientific Council on Radio Astronomy, Academician V. A. Kotel'nikov, it was possible to bring the laboratory of I. V. Shavkovskii and I. S. Rabinovich into the project in connection with building sensitive radiometers at centimetre wavelengths. By 1961, with their help and under the supervision of Salomonovich and Kuz'min, the groups of V. P. Bibinova and V. V. Ermolaeva had devised radiometers with parametric amplifiers operating at wavelengths of 1.6, 3.3 and 10 cm, which yielded an increase in sensitivity by about an order of magnitude. An apparatus designed to operate at 8 mm was developed at the Institute of Radio Physics and Electronics (V. A. Ablyazov, B. G. Kutuza), while a radiometer operating at 4 mm was devised at the Radio Physical Research Institute in Gorkii (Kislyakov).

In connection with the desire to conduct observations simultaneously (or nearly simultaneously) at several different frequencies, for example, to investigate the radio spectrum of Venus, a composite feed with electrical axes intended for simultaneous observations at 8 mm (in both polarisations) and at 1.6, 3.3, 10 and 21 cm was designed, manufactured and installed on the RT-22 (L. D. Bakhrakh, I. V. Vavilova, G. K. Galimov, Kalachev, A. M. Karachun, Kuz'min, Losovskii, Salomonovich).

As a result of a series of measurements by Yu. N. Vetukhnovskaya, Kislyakov, Kuz'min, Kutuza, Losovskii and Salomonovich, it was possible to obtain the spectrum of the brightness temperature of Venus from 4 mm to 10 cm for the first time.

The very high brightness temperatures that were measured at centimetre wavelengths did not necessarily imply that the surface and lower atmosphere of the planet had such high temperatures—a question that very much worried the engineers designing the equipment intended to land on the Venusian surface. Hoping to distinguish between the various models for Venus, two groups of researchers measured the radio brightness distribution over the disk of the planet in 1962. The Pulkovo group found a modest limb darkening toward the edge of the disk, consistent with the planetary surface being hot. However, a group of American radio astronomers obtained the opposite result.

In connection with the ambiguity of these interpretations of the spectrum and the associated radio brightness distribution, which led to very different conclusions about the physical conditions on Venus, Kuz'min proposed an experiment designed to identify the radiating medium, choose the appropriate model and thereby determine the temperature of the planetary surface. The idea behind the experiment was that, due to the differences in the Fresnel reflection coefficients for the vertical and horizontal polarisations, the radio emission of Venus should be polarised at the limb of the visible disk if this emission was radiated by the surface, but unpolarised if the emission was generated in the ionosphere, a cloudy layer or some other diffuse formation. This experiment, which was carried out in 1964 by Kuz'min and B. Clark on the radio interferometer of the California Institute of Technology, showed that the radio emission was polarised, thus demonstrating that the source of this emission was the surface of Venus, which had a temperature of about 650 K. The radius of the planet was also determined for the first time. These data were used in the development of the *Venera* landers, whose design incorporated thermal protection and a parachute of thermally stable fabric. In turn, the direct measurements of the temperature of the surface of Venus carried out in 1970–1972 by the *Venera-7* and *Venera-8* spacecraft confirmed that Venus is, indeed, a hot planet.

The equipping of the RT-22 with a very sensitive set of radiometers in the early 1960s, apart from various planetary measurements carried out by Kuz'min, Kutuza, Losovskii and Salomonovich (including the first observations of the 8 mm radio emission of Jupiter, Mars, Saturn and Mercury in the USSR), made it possible to conduct important studies of other cosmic radio sources. For example, in early 1963, V. A. Udal'tsov conducted polarisation studies of the Crab Nebula at 3.3 and 1.6 cm using the RT-22. These observations yielded the first measurement of the rotation measure in the direction of the Crab Nebula ($a = 0.28 \times 10^{-2}$ rad/cm^2). Udal'tsov obtained data indicating that there were different spatial distributions for the relativistic electrons giving rise to the optical and radio emission of the Crab Nebula.

The first international radio astronomy programme on the territory of the Soviet Union was carried out on the RT-22 telescope using the 8 mm radiometer in 1964. Kutuza, Matveenko, Salomonovich and the American radio astronomer A. Barret obtained the two-dimensional radio brightness distribution of the Crab Nebula at 3.3 and 0.8 cm. This same equipment was used for the detection of high-frequency excesses in the spectra of a number of quasars.

In the 1960s and 1970s, the development of work on the RT-22 was associated primarily with spectral studies of interstellar, monochromatic radiation and the installation of quantum paramagnetic amplifiers (QPAs) on the radio telescope.

At the beginning of 1963, Sorochenko, who had completed a series of investigations of the 21-cm hydrogen line at the Crimean Station, proposed to carry out a search for and study of radio lines of excited hydrogen on the RT-22. According to his calculations, searches for these lines, which had been predicted earlier by Kardashev, would be most effective at wavelengths near 3 cm, where the capabilities of the RT-22 matched the requirements of this task very well.

In order to realise this programme, Sorochenko and E. V. Borodzich, aided by M. T. and L. A. Levchenko, devised a new zero radiospectrometer intended for the search for radio lines, based on the 3-cm radiometer referred to above. This radiospectrometer was installed on the RT-22 telescope in early 1964. On the first day of observations, April 27, 1964, Sorochenko and Borodzich were able to detect a radio line of excited hydrogen in the direction of the Omega Nebula. This line, which was clearly visible even in individual recordings, corresponded to transitions between the 91st and 90th levels of the energetic states of the hydrogen atom. Subsequent measurements confirmed the results of these first observations.

The State Committee for Inventions and Discoveries registered the radiation of radio lines by excited hydrogen as a discovery made by FIAN researchers R. L. Sorochenko, E. V. Borodzich and N. S. Kardashev (who had first predicted this phenomenon), together with the husband and wife radio astronomers of the Pulkovo Observatory A. F. and Z. V. Dravskikh, who had carried out analogous measurements of radio lines at 5 cm. The date of the discovery was indicated as August 31, 1964, when a report on the detected lines was given at the XII IAU General Assembly.

In 1963, at the Oscillation Laboratory, under the supervision of Prokhorov, his PhD student R. M. Martirosyan built a QPA operating at a wavelength of 21 cm, which was successfully used on the RT-22 as part of a radiospectrometer complex (Martirosyan, Prokhorov, Sorochenko). This was the first such amplifier in the USSR constructed for radio astronomy studies. Its appearance proved to be very timely in other connections as well.

On April 16, 1964, it was possible to observe an occultation of the Crab Nebula by the Moon from Pushchino. It proved to be convenient to observe this very rare event at 21 cm: the signals from the Crab Nebula and the Moon were nearly identical for observations on the RT-22. Pushchino observations at 3, 10 and 21 cm yielded unique data on the brightness distribution of the Crab Nebula radio emission (Matveenko, Martirosyan, Sorochenko). Analogous observations were carried out at 8 and 32 cm on the antennas of the Deep Space Communications Centre in the Crimea, and also at metre wavelengths at Pushchino.

Work with the QPAs operating at 21 cm demonstrated that these amplifiers provided fundamentally new possibilities for radio astronomy studies, especially for observations of spectral lines.

In 1964, at the initiative of Salomonovich, radio astronomers of FIAN established close contact with the group of V. B. Shteinshleiger, who worked on the development of industrial models of the QPAs. Due to their joint efforts, a number of high-sensitivity radiometers with QPAs were devised for the RT-22 telescope. The development of the QPA radiometers was conducted under the supervision of Sorochenko.

A new group focusing on studies of the recently discovered radio lines of excited hydrogen and work on the development of new instruments for the RT-22 was formed, with its members being V. I. Ariskin, E. B. Borodzich, I. I. Berulis, V. M. Gudnov, L. M. Nagornykh, Yu. G. Chugunov and others. In their work, this group relied on the experience of N. F. Il'in, S. K. Palamarchuk and V. I. Pushkarev.

A radiospectrometer with a parametric amplifier was devised in 1966 for studies of the 21-cm neutral-hydrogen line on the RT-22 radio telescope (Berulis, B. Z. Kanevskii, A. A. Spangenberg, I. A. Strukov). A new zero radiospectrometer with a QPA operating at 5.2 cm was introduced into action in the same year. The system noise temperature was 130 K. The use of symmetric beam modulation with switching of the receiver input between two identical horn feeds made it possible to substantially reduce the influence of atmospheric fluctuations, which was important when using the high-sensitivity QPA radiometer. This radiospectrometer was used to carry out a survey of HII regions in the H 104 line of excited hydrogen, with detections being obtained in eight sources.

In 1964, anomalous high-frequency excesses were detected in the radio spectra of several extragalactic radio sources, due, as it was later elucidated, to the presence of components with very small angular sizes (10^{-2}–10^{-3} arcmin) in the nuclear regions of these galaxies and quasars. One of the earliest studies dedicated to the spectra and then the variability of the core components of active galaxies at centimetre and millimetre wavelengths was carried out on the RT-22 telescope under the supervision of Matveenko. V. I. Kostenko also actively participated in these studies.

Since the capabilities of the RT-22 telescope could best be applied in millimetre observations, it was always considered important to equip the telescope with sensitive equipment at these wavelengths. In 1964, nearly simultaneous with the development of the radiometer and 5.2 cm QPA, a similar radiometer for 8 mm observations was built. Since it was not possible to place the QPA at the prime focus of the telescope due to the operating conditions, and the losses in the millimetre waveguides were very high, Bakhrakh and Grigor'eva together with Kuz'min and Matveenko developed a two-mirror Cassegrain system at the initiative of Salomonovich, essentially simultaneous with its construction. This system made it possible to lower the antenna noise and fully realise the high sensitivity of the QPA. The Institute of Radio Physics and Electronics (V. A. Puzanov) also became involved in this work carried out under the supervision of Sorochenko and Kuz'min, who collaborated on the construction of a spectral radiometer operating at 8 mm wavelength that was used to study a QPA devised under the supervision of Shteinshleiger. All these developments brought about a sharp increase in the efficiency of the RT-22 telescope at 8 mm, by more than an order of magnitude.

A new system operating at 36,466 MHz was used to detect the H 56α radio line of excited hydrogen in 1968 (Sorochenko, Puzanov, Salomonovich, Shteinshleiger). This was the first radio line detected in the millimetre range, which later proved to be so fruitful for spectral investigations. Puzanov, K. S. Stankevich and Salomonovich measured the cosmic background radiation with this same radiometer (without the QPA) in 1967.

8 mm became the main operational wavelength for the RT-22 telescope. It was at this wavelength that Berulis and Sorochenko carried out a survey of thermal radio sources, while Kuz'min, Losovskii and Vetukhnovskaya conducted studies of planets, including the first detection of the radio emission of Uranus in the USSR. Measurements of the brightness temperature of Venus over its full phase cycle showed that day-to-night variations of the atmospheric temperature of Venus do not exceed 1%. The electrical parameters of the Martian surface were determined. In addition, the first detection in the USSR of radio emission from a planetary companion—Jupiter's moon Callisto—was made.

In subsequent years, a spectral radiometer with a 13.5 mm QPA was built. The importance of observing at this wavelength was determined by the discovery of maser emission by H_2O molecules, which opened the possibility of observations using interferometers with very long baselines.

The RT-22 telescopes of FIAN in Pushchino and of the Crimean Astrophysical Observatory in Simeiz were used together with an equipment complex developed at the Space Research Institute (IKI) to realise a Very Long Baseline Interferometry (VLBI) experiment (Matveenko, L. R. Kogan, Kostenko, I. G. Moiseev, Sorochenko, L. S. Chesalin). The possibility of devising a radio interferometer with independent recording of signals at the telescopes was first demonstrated by FIAN researcher L. I. Matveenko, who reported on this technique at a seminar at the Radio Astronomy Station in Pushchino in Autumn 1962. The first publication in connection with this question was somewhat delayed, and appeared only in 1965.

In 1969, the director of FIAN D. V. Skobel'tsin and the director of IKI G. I. Petrov decided to collaborate on studies of VLBI methods, which Matveenko was assigned to supervise.

As part of attempts to enhance the sensitivity of VLBI observations, a radiometer operating at 8 mm based on a travelling waveguide with a transmission bandwidth of 2000 MHz was tested. This radiometer was used in August 1971 to measure the radio emission of Mars during a grand opposition. The simplicity of its use gave preference to the travelling waveguide radiometer over the QPA radiometer for observations that did not require spectral resolution.

Investigations Using the DKR-1000 Telescope and Radio-Relay-Linked Interferometer Scientific observations on the DKR-1000 radio telescope began in 1964. The main focus of these first studies was constructing the spectra of a large number of radio galaxies and quasars in the relatively poorly studied metre-wavelength range. Artyukh, Vitkevich and Dagkesamanskii obtained the first measurements of the flux densities of radio sources at 38 and 86 MHz, using two correlation radiometers constructed in 1963 by Vlasov, Gavrilenko and Medvedeva.

Analogous measurements at 60 MHz were acquired in 1966, with the participation of colleagues from the Byurakan Astronomical Observatory. The extensive new observational material obtained made it possible to carry out a statistical analysis of the spectral characteristic of several hundred extragalactic objects, and to estimate the magnetic-field strengths and electron densities in individual radio sources. Dagkesamanskii detected a variation in the mean spectral indices of quasars with redshift (or equivalently, with the age of the quasars).

In 1965, T. D. Antonova (Shishova), Vitkevich and Vlasov carried out scintillation observations of the quasars 3C48, 3C144, 3C147 and 3C196 at 36 and 86 MHz on the V-3 antenna. Two correlation-type receiver complexes with orthogonal outputs were constructed in 1967, for use with the cross radio telescope at frequencies of 86–110 MHz (I. A. Alekseev, V. V. Gudnova, Yu. F. Sigaev). These receivers proved to be long-lived, and were subsequently used for a large number of scientific programmes.

Transistor-based antenna pre-amplifiers were constructed in 1965–1967 to enhance the sensitivity of the V-3 antennas of the DKR-1000 radio telescope (A. N. Cheremisin, Yu. P. Shitov, P. P. Nikolaev, S. T. Nuzhdin). A system of 18 amplifiers phased over a wide bandwidth (60–120 MHz) was installed on the V-3 antenna, increasing the sensitivity of the radio telescope by approximately an order of magnitude.

In 1968, Ilyasov and S. N. Ivanov carried out observations of the eclipse of the Crab Nebula by the solar supercorona under the supervision of Vitkevich, using the T-shaped DKR-1000 radio telescope with the 2PR receiver. The first such observations of the Crab Nebula at metre wavelengths were obtained with a pencil beam $10 \times 13'$ in size.

In 1970, Udal'tsov, V. N. Brezgunov, A. N. Tolstov and I. I. Galyamova devised an eight-channel receiver with a frequency range of 60–115 MHz (with the width of each channel being 120 kHz), designed for studies of variations of pulsar spectra.

It was already clear in the 1960s that studies of the structure of discrete sources required high angular resolution, better than $1'$. At the initiative of Vitkevich, work began on the creation of a radio interferometer based on the DKR-1000 telescope and a number of comparatively small antennas forming baselines of several tens of kilometres, with the received signals to be conveyed to a central processing point along a radio-relay line.

Work on this interferometer moved forward after 1965, when G. I. Dobysh and a group of researchers in the Radio Technical Department at the Tula Polytechnic Institute began to manufacture the required radio apparatus. Toward the Summer of 1968, Dobysh and the young FIAN engineers working with him B. I. Ivanov, M. T. Rezepin and V. A. Frolov, devised a radio interferometer designed to operate at 35 MHz, and carried out the first interferometric observations of several of the most powerful radio sources using the V-3 antenna and a simple movable antenna forming a baseline of 13 km.

In October 1971, under the supervision of Vitkevich and Dobysh, researchers of the Radio Astronomy Station (B. K. Izvekov, S. A. Sukhodol'skii, V. A. Frolov) completed work on the construction of a new radio interferometer with a variable baseline designed to operate at 85 MHz. The parameters of the radio-relay line enabled operation with baselines up to 40 km in length. The V-3 antenna of the DKR-1000 telescope was used as the main element in this interferometer. The antenna used for the movable site was an easily transportable array made up of 64 waveguide elements with an effective area of about 200 m^2. This array was developed and constructed by Ilyasov, Solodkov, Tyaptin and V. Ya. Shcherbinin. Dagkesamanskii and a group of his co-workers used this interferometer to study the structures of about 150 radio sources with an angular resolution of about 30″.

In 1964, Vitkevich proposed to study the solar wind using the DKR-1000 telescope and two new antennas to be placed at distances of 150–250 km from Pushchino. It was proposed to measure the time shifts between the scintillation patterns obtained at each site, and thereby to measure the speed of plasma inhomogeneities in the solar supercorona. Under the general supervision of Vitkevich, two movable antennas were constructed near Staritsa in the Kalinin Region and in Pereslavl'–Zalesskii in the Yaroslav Region in 1965–1966 (Afon'kin, V. A. Egorov, Ilyasov, S. M. Kutuzov, Tyaptin).

The equipment for these sites was devised by Alekseev, Gudnova, M. V. Sil'yanov and Sigaev. The most outstanding feature of these antennas was their ability to operate simultaneously at two wavelengths (3.5 and 7.0 m). The parabolic cylindrical antennas were 280 × 20 m in size, and were placed low above the ground. The first measurements of the speed of the solar wind using radio astronomy methods in the USSR were carried out using this system in 1966.

The results of all these studies laid the basis for a report on investigations of near-solar space using radio astronomy methods carried out by radio astronomers of FIAN made by Vitkevich in 1966 at a session of the Presidium of the USSR Academy of Sciences, which was met with considerable interest by M. V. Keldysh and members of the Presidium. Vitkevich's series of investigations of the supercorona, including observations of the solar wind, was awarded a State Prize of the USSR in 1969—the first State Prize received by a scientist working in the area of radio astronomy.

In 1969, the movable antenna at Pereslavl'–Zalesskii was used together with the V-3 antenna of the DKR-1000 telescope in the first Soviet VLBI experiment carried out by the Radio Physical Research Institute in Gorkii and the Radio Astronomy Station of FIAN (Alekseev, Vitkevich, E. D. Gatelyuk, Troitskii and others).

In early 1968, English radio astronomers announced the discovery of a new class of object—pulsars. Studies of pulsars began at the Radio Astronomy Station immediately after this report, at the initiative of Vitkevich. Appropriate receiver–detection equipment was devised in short order. A two-channel radiometer based on R-313 receivers and operating at two adjustable frequencies in the range 60–120 MHz was constructed in 1968. This apparatus was used to carry out the first studies of the pulsars SR 1919 and SR 0809 (Alekseev, V. F. Zhuravlev and Shitov).

A 12-channel spectral analyser operating at 60–120 MHz was developed in 1969, intended to enable studies of the fine structure of pulsar spectra due to scintillations in the interstellar plasma. This analyser was used to measure the polarisation and strength of the Galactic magnetic field in the interstellar medium in the directions of a number of pulsars.

In 1968, Vitkevich and Shitov discovered that the pulses of the pulsar PSR 0809+74 consisting of two or three individual subpulses (components) regularly drift relative to the main mean period of the pulsar, appearing at an earlier phase in each new period and disappearing before the leading edge of the "window" for the pulsar's radio emission.

In that same year, Vitkevich, Yu. I. Alekseev, Zhuravlev and Shitov discovered one of the first new pulsars, named PP 0943: Pushchino pulsar with right ascension 09 hr 43 min. Vitkevich made a presentation about the results of these pulsar studies at a session of the Presidium of the USSR Academy of Sciences on June 11, 1970. In discussions with scientists promoting various pulsar models, Vitkevich defended the "rotating lighthouse" model proposed by himself earlier, which indeed proved to be physically correct. A large group of workers in the Radio Astronomy Laboratory were awarded a prize of the Presidium of the USSR Academy of Sciences for their investigations of pulsars.

In connection with the high degree of polarisation of the radio emission of pulsars, which provided additional information about the mechanism generating this radio emission, it was of interest to obtain measurements in two orthogonal polarisations.

In 1968, Ilyasov, Kutuzov and Tyaptin developed a simple antenna array that received radiation polarised orthogonal to the polarisation of the DKR-1000 telescope at 3.5 m. Simultaneously, Yu. I. Alekseev designed a polarimeter and developed a method for calibrating the polarisation of an interferometer consisting of the V-3 antenna of the DKR-1000 telescope and the new "polarisation" antenna. A small additional antenna that could receive radiation polarised at an angle of 45° to the polarisation of this antenna was built for the purposes of this calibration. Most of the observations on the polarisation antenna were carried out by Alekseev and S. A. Suleimanova.

In 1969, Matveenko and Lotova showed that the observed angular size of the pulsar in the Crab Nebula is determined by scattering, and should increase in proportion to the square of the wavelength, while the duration of the pulses should increase in proportion to the fourth power of the wavelength.

The Large Scanning Antenna of FIAN These pulsar studies led to the creation of a third large-scale radio telescope—the Large Scanning Antenna (BSA) of FIAN. Vitkevich was the first to point out that radio telescopes intended for pulsar studies can be constructed according to a filled-aperture scheme, but being aware to within known limits of restrictions on their sensitivity due to the effect of confusion. Together with Ilyasov, he started the construction of a large metre-wavelength radio telescope in the form of a large phased array. A team working under the direction of Vitkevich (Udal'tsov, Alekseev, Shitov, Grezgunov, Kutuzov, Tyaptin, K. K. Darkov) took on the task of planning the corresponding antenna–equipment complex for pulsar investigations.

The manufacture and construction of the new radio telescope began in the beginning of 1969, and the elements were assembled in 1970–1972. The main constructional and technical directors were Kutuzov and Tyaptin. This was the main work of the laboratory (the assistant heads of the laboratory at that time were Udal'tsov and N. P. Vladimirov). Nearly all the engineers in the laboratory were employed to carry out this work.

The combination of the proposed temporal methods for phasing (a discrete phase rotator) and phase methods (a beam-forming matrix) appreciably simplified the microwave system, and turned the BSA radio telescope into an instrument that was able to realise operative changes in the direction of signal reception and a multibeam antenna beam (Ilyasov, A. A. Glushaev, S. N. Ivanov, Kutuzov, Sodolkov and others).

The efficiency of the new single-frequency radio telescope depended on its stability against interference. A two-stage system for distributing the amplification was employed to protect the antenna amplifiers from cross interference (television and ultra-short-wave FM stations), and narrow-band filters were installed at the amplifier inputs (G. F. Novozhenov, I. A. Alekseev and others).

V. V. Vitkevich died in the prime of life in January 1972. His passing represented a heavy loss for Soviet radio astronomy.

The general direction of the work was taken on by Kuz'min, who was the new head of the laboratory and director of the Radio Astronomy Station. The scientific supervision for the antenna–equipment complex for the BSABSA was entrusted to Ilyasov, who was named assistant head of the laboratory. A large amount of help in the final stage of the work was given by the assistant director of FIAN S. I. Nikol'skii. The construction of the Large Scanning Antenna of FIAN was completed in 1974 (Fig. 1.17).

In 1976, at the initiative of Udal'tsov and under his supervision, the BSA was linked with the V-3 antenna of the DKR-1000 telescope in some observations, in order to enhance the resulting spatial resolution of the system. This interferometer was used to conduct observations of several discrete radio sources with high resolution and sensitivity.

The BSA was used for studies of pulsars (Kuz'min, Shitov, Izvekova, I. F. Malov, V. M. Malofeev, T. V. Smirnova, Suleimanova, T. V. Shabanova), the interplanetary plasma (V. I. Shishov, Vlasov, T. D. Shishova), supernova remnants (Udal'tsov, A. P. Glushak, A. V. Pynzar'), clusters of galaxies (Dagkesamanskii) and other objects.

A survey of northern pulsars was carried out in 1974, and the first low-frequency catalogue of the fluxes and profiles of pulsar pulses, which included about 100 objects, was compiled. As a result of studies of pulsar spectra, a number of which were conducted in collaboration with the Institute of Radio Physics and Electronics and Jodrell Bank Observatory (England), it was established that a low-frequency "cut-off" is characteristic of the radio emission of pulsars, with the maximum emission usually being concentrated, on average, near a frequency of 100 MHz. It was discovered that, during the evolution of a pulsar, as its rotation slows down (its period increases), the radio spectrum and its maximum shift toward lower frequencies, and the magnetic axis approaches the rotational axis.

Fig. 1.17 BSA radio telescope. Antenna-feeder system

Investigations of the linear polarisation of pulsars were carried out using an original technique developed in the laboratory and based on the use of Faraday rotation in the interstellar medium. It was discovered that the variations of the degree and position angle of the polarisation along the mean pulsar profile at metre wavelengths differ substantially from the pattern observed at shorter wavelengths.

Four new pulsars in the solar neighbourhood were discovered, one of them (PSR 0320+39) displaying an anomalous (opposite to the usual direction), regular, highly organised subpulse drift.

Thanks to the high sensitivity of the BSA radio telescope and the application of specialised methods for processing of the observations, it was possible to achieve a time resolution of 10 μs. These studies, which were conducted jointly with the Space Research Institute, showed that the individual pulses of pulsars at metre wavelengths are made up of numerous micropulses. The simultaneous existence of two fine temporal structures with different periodicity time scales in pulsar pulses was discovered.

Udal'tsov, Glushak and Pynzar' detected radio halos around two pulsars, which apparently arose as a result of the interaction of relativistic particles ejected by the pulsar with the interstellar medium.

Observations of pulsars on radio telescopes and long-baseline interferometers required the installation of a high-accuracy local time standard at the Radio Astronomy Station, which would provide a time signal synchronised with the USSR state reference time. The realisation of this system was carried out by A. S. Vdovin under the supervision of Ilyasov, with active methodical help from the All-Union Scientific Research Institute for Physical and Radio Technical Standard Measurements (V. G. Il'in, A. R. Oksentyuk, G. N. Palii, S. B. Pushkin).

The ability to rapidly redirect the BSA beam made it possible to observe up to 500 scintillating sources in the course of a day. Such measurements, proposed and carried out by Vlasov in 1975–1980, were used to determine the time variations of the large-scale structure of the interplanetary plasma, and to investigate their relationship to solar activity and geomagnetic phenomena. It was shown that, at the period of maximum solar activity (1979), the interplanetary plasma was distributed nearly spherically symmetrically, while it was appreciably elongated along the plane of the solar equator in the period of minimum solar activity (1976).

The structures of extragalactic radio sources were studied using the BSA and the scintillation method. The results were interpreted to indicate that quasars are not a special class of object, and instead are radio galaxies that are located in a stage of very high core activity. Artyukh and Vetukhnovskaya used scintillation observations to study the structures of more than a hundred peculiar extragalactic objects: BL Lac objects, quasars, Seyfert galaxies, Markarian galaxies, etc. (More recently, colleagues at the Byurakan Astronomical Observatory have also taken part in this programme.)

In addition to experimental work at the Radio Astronomy Station, a series of theoretical studies related to the passage of radio waves through the interplanetary plasma was conducted. V. I. Shishov developed a correlation theory for strong scintillations, introduced the main equation of this theory and analysed solutions to this equation.

In 1967–1969, N. A. Lotova and A. A. Rukhadze (Plasma Physics Laboratory of FIAN) laid out the basis for the theory of turbulence in the solar wind. They introduced concepts relating inhomogeneities in the electron density giving rise to scintillations with wave processes, and considered the development of instability in the interplanetary medium.

In 1973–1981, Lotova, I. V. Shachei, D. F. Blums and other researchers at the Radio Astronomy Station carried out a series of studies dedicated to the theory of the formation of a non-stationary scintillation pattern.

In 1981–1982, Lotova, Sorochenko and Blums proposed and successfully applied a new method for studies of the solar wind—scintillation observations of water maser sources. M. V. Konyukov solved the very important problem of reconstructing the phase front of a wave that is distorted during propagation through an inhomogeneous medium (such as the ionosphere). This opened possibilities for creating large aperture-synthesis systems operating at metre and decimetre wavelengths.

Modernisation and Automation of the Radio Astronomy Station The development of the experimental base at the Radio Astronomy Station was facilitated by the continuous modernisation of its unique radio telescopes. As early as January 1965, in response to a proposal from FIAN, a special commission of the Presidium of the USSR Academy of Sciences made a decision on the main directions for modernisation of the FIAN radio telescopes.

The development and realisation of projects to modernise the DKR-1000 and RT-22 radio telescopes were carried out by the Radio Telescope Department (P. D.

Kalachev, V. P. Nazarov, I. A. Emel'yanov, V. L. Shubeko, G. A. Pavlov and others), the Spectral Radio Astronomy Department (R. L. Sorochenko, I. I. Berulis, V. A. Gusev, G. T. Sirnov, A. M. Tolmachev and others), the Antenna Group (Yu. P. Ilyasov, S. N. Ivanov, V. T. Solodkov and others), the Equipment Group (G. N. Novozhenov, Yu. I. Alekseev, G. I. Dobysh, I. A. Alekseev and others), the Electronic–Computational Device Group (Yu. V. Volodin, A. A. Sal'nikov and others), and the Groups for Exploitation of the DKR-1000 (P. D. Tsyganov, V. M. Karlov, V. V. Ivanova and others) and the RT-22 (L. M. Nagornykh and others).

Major work on the modernisation of the control system for the north–south antenna beam of the DKR-1000 cross radio telescope was completed in 1978. An antenna-beam control system capable of tracking a source in a 4° sector was developed in order to enable a tracking regime for the V-3 antenna of the DKR-1000 telescope. This system could be implemented at 30–120 MHz without changing the electrical length of the antenna system, which is especially important when using the V-3 antenna in interferometric observations or joint observations with the north–south antenna of the DKR-1000.

The receiver apparatus for the radio telescope was devised at the Radio Electronics Department of the Tula Polytechnic Institute under the supervision of V. V. Davydov. A large contribution was made by Dobysh, who later transferred from the Tula Polytechnic Institute to the Radio Astronomy Station of FIAN, and was named assistant head of the laboratory in 1979. A three-channel receiver was brought into use in 1981.

Automation of the DKR-1000 was based on the use of an M-6000 computer, and was initially developed at the Automation Department of the Tula Polytechnic Institute, with the participation of Udal'tsov and Artyukh (Fig. 1.18). In the initial stages, a large role in the automation of the data reduction was played by the installation of multi-channel recording, realised in 1972 at the Institute of Radio Physics and Electronics for use at the Radio Astronomy Station. Filter spectral analysers for pulsar studies were devised at the end of the 1970s. A 16-channel analyser with a frequency bandwidth of 160 kHz per channel was introduced into use in 1977 (Alekseev and others). A 128-channel with a channel bandwidth of 20 kHz was manufactured in 1979 (Dobysh and others).

The programme for modernisation of the RT-22 radio telescope foresaw the creation of an automated system for the control of the radio telescope and the collection, accumulation and initial processing of the observational data.

In 1970, Sorochenko and Udal'tsov suggested the use of a series of industrially produced computers in the automation systems for radio astronomy observations on the RT-22 and DKR-1000 radio telescopes. In 1971, at the initiative of Sorochenko, the Division of Computational Devices of FIAN (headed by A. V. Kutsenko) worked on a system for the automation of the RT-22 based on an M-6000 computer. One fundamental property of this project developed jointly with this division was its comprehensive nature. The computer controlled the entire operation of the radio astronomy experiment: the pointing of the telescope, control of the receiver, and the collection and reduction of the observational data. The realisation of this project in its entirety required more than five years.

Fig. 1.18 Members of the Scientific Council of the USSR Academy of Sciences on Radio Astronomy (Pushchino, December 1981) get acquainted with the control system of the DKR-1000. *Right to left*: Yu. N. Pariiskii, K. s. Shcheglov, V. A. Kotel'nikov, A. D. Kuz'min, R. D. Dagkesamanskii, S. Ya. Braude. V. M. Malofeev is at the controls

The first system with a TPA/I mini-computer was introduced on the RT-22 in January 1974, when it was used during observations of Comet Kohoutek.

A large contribution to the development of the automation of the RT-22 drive system was made by Kuz'min, who attracted the group who prepared the first analog system for the driving and control of the RT-22 to this work (G. N. Posokhin, B. K. Chemodanov, V. A. Vvedenskii, Yu. N. Semenov).

Kalachev and his co-workers developed a special design for sensor equipment designed to eliminate the influence of backlashes in the radio telescope drive and perpendicular shifts of the rotating platform. The final accuracy of the pointing was verified using a specially developed programme that enabled the computer to use selected guide sources to find corrections to the pointing and take them into account; in other words, the programme implemented a certain "self-teaching" of the computer. A project pointing accuracy of $10''$ was realised by the RT-22 pointing system. Researchers in the division of Kutsenko (E. A. Zubova, B. A. Polos'yants, S. A. Terekhin, V. A. Shirochenkov) actively took part in the automation of the RT-22 together with their colleagues at the Radio Astronomy Station. The scientific and technical supervision of the construction of the automation complex for the RT-22 was carried out by Sorochenko and Kutsenko.

This system was introduced into use on October 5, 1978. The efficiency of using the radio telescope was increased by an appreciable factor, making it possible to substantially broaden the programme of observations conducted on the RT-22 telescope. The now traditional studies of radio lines of excited hydrogen at 8 and 13 mm were joined by observations of molecular radio lines. Radiation corresponding to the $J = 4 \to 3$ rotational transition of the cyanacetylene molecule was detected, and these measurements used to derive the physical conditions in a number of molecular clouds in which star-forming processes were occurring (Sorochenko, Tolmachev).

A programme of systematic observations of a large number (more than 60) of H_2O maser sources was conducted in collaboration with colleagues at the Sternberg Astronomical Institute, with the aim of elucidating the regularities in their variability (Sorochenko, Berulis, E. E. Lekht, G. M. Rudnitskii, M. I. Pashchenko).

The current Radio Astronomy Station of FIAN is a major radio astronomy centre that is well known for its scientific and technical work both in the USSR and abroad. Two large-scale metre-wavelength radio telescopes (the DKR-1000 and LSA) operate at the Radio Astronomy Station, as well as one of the largest millimetre-wavelength radio telescopes (the RT-22). The scientific topics covered by the research carried out at the station include radio astronomy investigations of the Sun, planets, the interstellar and interplanetary media, pulsars, supernova remnants, and quasars, radio galaxies and other galaxies in both their continuum and spectral lines.

Researchers at the Radio Astronomy Station have carried out a large amount of work on the creation and implementation of various methods for the reception of weak signals from cosmic radio sources, original and unique receiver equipment, automation systems and a whole series of specialised radio technical schemes and systems that have found employment in both radio astronomy and in various other technical areas.

Radio astronomers of FIAN conduct collaborative investigations with scientists at the Space Research Institute, Radio Physical Research Institute, Institute of Radio Physics and Electronics, Byurakan Astronomical Observatory, Leningrad State University and the Sternberg Astronomical Institute. Long-term scientific collaborations have been established between the Radio Astronomy Station and the Jodrell Bank Observatory (England), the Max Planck Institute for Radio Astronomy (Federal Republic of Germany) and the University of Tasmania (Australia). Many foreign scientists have come to the Radio Astronomy Station of FIAN to conduct observations using its radio telescopes.

1.4 Radio Astronomy Studies in the Theoretical Physics Division

The beginning of work on radio astronomy carried out in the Theoretical Physics Division of FIAN can be taken to be the calculations published by V. L. Ginzburg in his paper *On the Radiation of the Sun at Radio Frequencies* (1946), which played a determining role in our understanding of the nature of solar radio emission. Theoretical radio astronomy studies gradually developed in subsequent years, with a central position being occupied by theoretical analyses of radiation mechanisms, the propagation of waves and relativistic particles and a number of problems in Galactic, extragalactic and solar radio astronomy.

S. I. Syrovatskii (Fig. 1.19), who came to the Theoretical Physics Division in 1951, actively took part in these studies. The death of Syrovatskii on September 26, 1979 broke off his scientific activity when he was at a high peak in his abilities.

Fig. 1.19 Sergei Ivanovich Syrovatskii (1925–1979)

Theory of Cosmic Synchrotron Radiation At the end of the 1940s and the beginning of the 1950s, the concept that non-solar radio emission, which was predominantly non-thermal, was associated with some class of cosmic radio source—"radio stars"—was widespread. The alternative hypothesis that non-thermal cosmic radio emission was synchrotron radiation, which had appeared in the scientific literature outside the Soviet Union in 1950, long remained in shadow. In 1951, Ginzburg showed that synchrotron radiation by relativistic electrons in Galactic magnetic fields "is very natural and attractive as an explanation for the general radio emission of the Galaxy." Ginzburg analysed the relation between the flux of the electron component of the cosmic rays and the intensity of this component's synchrotron radiation. In their subsequent papers, Ginzburg and Syrovatskii showed how it was possible to study various characteristics of Galactic relativistic electrons based on radio astronomy observations of non-thermal Galactic radio emission, and, in some cases, the characteristics of cosmic-ray protons and nuclei in our own and other galaxies. The radio astronomy theory of the origin of cosmic rays that was developed on the basis of these studies now occupies an important place in modern astrophysics. Moreover, the establishment of the relationship between radio astronomy and the physics of cosmic rays led to the birth of a new direction in astronomy—the astrophysics of cosmic rays, and then high-energy astrophysics.

During the development of these areas, fundamental questions in the theory of synchrotron radiation were posed and solved, having to do with the spectral and polarisation characteristics of the radiation in the case of non-circular motion of the radiating electrons, the influence of reabsorption with a power-law spectrum, etc. (Ginzburg, Syrovatskii, V. N. Sazonov). The monograph of Ginzburg and Syrovatskii *The Origin of Cosmic Rays* (Moscow, Leningrad; USSR Academy of Sciences; 1963) is still widely used today as a scientific textbook. This monograph was again published in 1964 after having been supplemented and translated into English.

Theory of Radio Galaxies, Quasars and Their Variable Radio Emission With the discovery of radio galaxies, and then quasars and activity in the nuclei of various types of galaxies, the nature of the energy release in these objects became a burning question, which has remained one of the central problems in astrophysics right to the present day. Ginzburg related this energy with the release of gravitational energy (for example, associated with gravitational instability of the galaxy or its central regions). A number of fundamental studies were carried out in this direction, which retain their importance to this day.

In 1964–1966, Ginzburg and L. M. Ozernoi showed that the compression of a magnetised gaseous cloud is accompanied by a huge amplification of the magnetic-field strength. The compression of a massive, rotating cloud with a magnetic field leads to the formation of a magnetised plasma body—a magnetoid, which could serve as a source of activity in a quasar or galactic nucleus over a long time. Depending on its entropy, a magnetoid could be hot (quasi-spherical) or cool (in the form of a disk). Its structure, evolution and electrodynamical properties and the character of its radiation were studied in detail by Ozernoi and V. V. Usov (1971–1973). In particular, powerful non-thermal radiation with its maximum at submillimetre wavelengths or the far infrared, as is the case for many quasars and active galaxies, can be explained by the magnetoid theory. Instability accompanied by twisting of the magnetic-field lines connecting the magnetoid and the surrounding plasma can lead to repeating magnetorotational outbursts and the ejection of beams and jets of relativistic electrons from the magnetoid. In 1970, Ozernoi and B. M. Somov calculated the parameters of such magnetorotational outbursts, which could explain the energetics and a number of properties of the radio variability of quasars.

At the end of the 1960s and the beginning of the 1970s, observational data indicating the presence of relativistic effects in the radio variability of quasars became available. In 1968, Ozernoi and Sazonov developed the theory of synchrotron radiation for a relativistically moving jet and relativistically expanding spherical sources. In 1969, they calculated in detail the spectrum, polarisation and time variations of a radio source having the form of two components moving from each other in opposite directions. This model or modified versions of it are still referenced today in connection with the interpretation of apparent superluminal motions in compact radio sources. In 1974, Ozernoi and L. E. Ulanovskii proposed a theory of radio variability in the case of motion of relativistic electrons in an external spatially inhomogeneous magnetic field, and used this theory to interpret a number of effects observed in the radio variability of quasars.

In the early 1980s, it became clear that a large fraction of the power emitted by many quasars and active galactic nuclei is radiated not only in the far infrared, but also in hard X-rays and gamma-rays. A number of studies have interpreted this high-energy radiation of quasars as a result of inverse Compton scattering on the relativistic electrons. The important role of this effect in quasars and the possibility of obtaining powerful gamma-ray radiation via this mechanism was noted in 1963 by Ginzburg, Ozernoi and Syrovatskii, using 3C273 as an example.

Radio Emission and the Nature of the Galactic Centre The centre of our Galaxy contains a pointlike radio source that also radiates in other energy bands, and in this sense qualitatively resembles the centres of active galaxies and quasars, although the level of its activity represents a negligibly small fraction of the activity of these more powerful objects. For example, the power of its radio emission is a factor of 10^{10}–10^{11} weaker than the radio emission of a typical quasar. Many theoreticians have proposed that there is a massive black hole at the Galactic centre, which represents a "dead" quasar, that is, a quasar with an extremely low accretion rate. However, a series of studies carried out in the Theoretical Physics Division have shown that the inevitable tidal disruption of stars surrounding the galactic nucleus by the black hole should give rise to a flow of gas onto the black hole, with an associated emission of radiation. This contradicts available X-ray observations if the mass of the black hole exceeds 1000 times the mass of the Sun (Ozernoi, V. I. Dokuchaev, V. G. Gurzadyan).

Radio interferometric measurements of the size and structure of the point-like radio source in the Galactic centre have demonstrated the important role of scattering of its radiation on inhomogeneities of the electron density. In 1977, Ozernoi and Shishov showed that the scattering region is located in the vicinity of the radio source itself. Later, they presented arguments suggesting that the scattering inhomogeneities are located in compact clouds of ionised gas surrounding the radio source. These clouds were subsequently discovered in 1979. Ozernoi and Shishov have proposed that the 0.511 MeV annihilation line observed in the direction toward the Galactic centre originates precisely in these clouds. The energy of an electron–positron wind from the central source is sharply decreased in these clouds due first to adiabatic and then to ionisational losses, after which annihilation becomes possible. Observations of scintillation of the radio flux from the pointlike source were used to estimate its intrinsic size, which suggests it is either a young pulsar or a moderate-mass black hole.

Models for the Propagation of Cosmic Rays and the Background Radio Emission of the Galaxy In 1953, Ginzburg, following S. V. Pikel'ner, used the concept of a halo of cosmic rays, having in mind an extensive region around the Galactic disk where cosmic rays, including relativistic electrons, could be confined over long times before escaping into intergalactic space. This work essentially predicted that the size of the radio halo should depend on the frequency, since the relativistic electrons occupy only part of the halo due to energy losses, and occupy a smaller volume the higher their energy. This frequency dependence of the size of the halo was subsequently detected in studies of the radio emission of other galaxies.

In 1972–1976, S. V. Bulanov, V. A. Dogel' and Syrovatskii carried out careful calculations of the propagation of relativistic electrons in the Galaxy, taking into account synchrotron and Compton losses, and analysed the spatial distribution of the Galactic radio background. Based on an analysis of observational data on the intensity of the radio background at various frequencies and on the energy dependence of the electron component of cosmic rays, they likewise concluded that the Galaxy has an extended halo.

Observations of galaxies using radio telescopes with high spatial resolution began to be made approximately in the middle of the 1970s. These observations revealed extended radio halos with thicknesses of several kiloparsecs[7] around the optical disks of several galaxies. Observations at several frequencies were used to investigate the motion of cosmic rays in the halos of these galaxies. The most popular models for the propagation of cosmic rays at that time were the diffusion model of Ginzburg and Syrovatskii and a convection model, in which cosmic rays are carried from the Galaxy by a Galactic wind. Analysis of the spectra of the various components of cosmic rays observed at the Earth did not enable unambiguous identification of either of these models as being preferable; the characteristic properties of the spectra could be explained using either model with an appropriate choice of parameters.

Ginzburg then noted that more definite conclusions could be drawn using data on the radio emission of galactic halos. Since the size of the radio halo at some frequency is determined by the mean-free path of an electron with the corresponding energy, and the energy dependences of this mean-free path are different in the diffusion and convection models, the observed variation in the size of the radio halo with frequency should also be different in the models. Studies of the halo radio emission of galaxies showed that the propagation of electrons in the halos can be better described using the diffusion model, with the effective diffusion coefficient $D \simeq 10^{29}$ cm^2/s (Dogel', Kovalenko, Prishchep).

The diffusion of charged particles in galactic space is determined by the spectrum of inhomogeneities of the magnetic field. An expression for the relative fluctuations of the halo radio intensity was obtained with the aim of determining this spectrum. In 1981, V. S. Ptuskin and G. V. Chibisov showed that the anisotropic part of the correlation function for the radio intensity could be used to obtain the magnitude of the regular magnetic field. Experimental studies of these concepts require observations of the radio background at frequencies no higher than a hundred Megahertz and with an angular resolution of better than 1'.

The question of where and how cosmic rays are accelerated is among the most important astrophysical problems. Acceleration processes can also be studied using radio observations, since the acceleration of charged particles is often accompanied by the emission of radiation (including radio radiation) as a consequence of various types of energy loss.

Ginzburg showed that the acceleration of particles in shock fronts entering the halo could be detected from the synchrotron radiation of the accelerated relativistic electrons. A model for the propagation of a shock front through the Galactic halo was developed, in which the distribution of the electrons depended only on the electron energy, as a consequence of energy losses. It turned out that the electrons could be accelerated under certain conditions. The presence of such acceleration could be detected by comparing the spectra of the radio emission from the vicinity of the shock front and the radio emission from regions without acceleration. It was shown that variations in the spectral index of the radio emission are associated with two effects: the acceleration of particles at the shock front and adiabatic losses in the region behind the front. For acceleration to occur, the efficiency of the scattering

---

[7] A parsec is approximately 3.08×10^{16} m—Translator.

in the vicinity of the front must be higher than the average efficiency of scattering in the Galaxy (Bulanov, Dogel', Kovalenko). Specific data on primary cosmic rays near the Earth suggest that acceleration in interstellar space is, in all likelihood, inefficient (Ginzburg, Ptuskin).

Active Processes. Radio Emission of the Sun, Neutron Stars and Pulsars In 1946, the existence of strong thermal radio emission from the outer regions of the solar corona was hypothesised, which subsequently found full experimental confirmation (see Sect. 1.1). Following this, Ginzburg and G. G. Getmantsev proposed in 1952 that the sources of enhanced radiation above sunspots had a synchrotron nature. This hypothesis proved to be very fruitful in connection with explaining the various components of the solar radio emission (Type IV bursts, microwave bursts, etc.).

In 1958, Ginzburg and V. V. Zheleznyakov proposed a theory for the generation of rapidly drifting Type III bursts, and also studied in detail the propagation in and emergence from the coronal plasma of electromagnetic waves. An analysis of the combined scattering of plasma waves laid the basis for a broad series of investigations of decay interactions of electromagnetic waves in a plasma. In 1968, immediately after the discovery of pulsars, Ginzburg, in collaboration with researchers at the Radio Physical Research Institute in Gorkii, considered a number of questions in the theory of the radio emission of pulsars, as well certain properties of neutron stars.

A new push was given to studies of active processes on the Sun by theoretical work of Syrovatskii begun in 1966, concerned with the behaviour of plasma in a strong magnetic field. He showed that a current sheet with a width much larger than its thickness could form with time in the vicinity of a zero point of the magnetic field in such a plasma. One of the frequent applications of this work was modelling the behaviour of plasma in the solar corona, where the energy density of the magnetic field is much larger than the energy density of the plasma. The formation of current sheets in the atmosphere of the Sun is associated with the appearance of active regions on the photosphere (regions whose magnetic fields are strong compared to the background fields), as a result of which there arises an electrical field along the force lines of the magnetic field.

Over many years of investigations carried out under the supervision of Syrovatskii at a number of institutions (FIAN, the Institute for Terrestrial Magnetism, the Ionosphere, and Radio-wave Propagation, the Institute of Applied Mathematics, the Institute of Applied Geophysics), various processes having to do with the formation, stability and decay of current sheets were studied. Under the conditions in the solar corona, the lifetimes of these structures are several hours. This leads to the accumulation in the solar atmosphere of colossal energies in the form of the magnetic energy of current sheets. In the model of Syrovatskii, the disruption of current sheets accompanied by the release of this accumulated energy in various forms is observed as a solar flare.

The parameters of a pre-flare current sheet should differ appreciably from those of the surrounding coronal plasma, and its spectrum should differ from the spectrum of the unperturbed solar atmosphere. In 1977, Syrovatskii suggested the possibility of predicting solar flares using radio astronomy methods. These and later studies carried out by Syrovatskii together with V. S. Kuznetsov showed that the radiation of the current sheet leads to an increase in the flux from the Sun in some range of radio wavelengths, while the current sheet simultaneously blocks the radiation of the lower solar atmosphere, leading to a decrease in the flux at certain other radio wavelengths. It was concluded that it should be possible to detect a pre-flare situation on the Sun using already existing radio telescopes with high resolution operating at centimetre and millimetre wavelengths.

New Methods for Radio Astronomy Investigations In 1947, Ginzburg suggested making observations of the diffraction of the radio emission from discrete sources at the limb of the Moon as a means of studying their structure (the method was developed jointly with Getmantsev in 1950). An opportunity to apply this high-resolution method arose when the recently discovered quasar 3C273 passed behind the Moon in 1963, making it possible to reveal the presence of complex structure.

In 1960, Ginzburg proposed a new method for determining the magnetic-field strength in the outer solar corona by observing the polarised radio emission from a background source that had passed through the coronal material. This was based on observations of the rotation of the plane of polarisation, as well as any depolarisation of the radio emission.

In 1956–1963, Ginzburg and V. V. Pisareva worked out a method for studying the inhomogeneous structure of the circumsolar plasma using observations of the diffractive scintillation of compact background radio sources due to inhomogeneities in the plasma. Such scintillations were detected in 1964, and were used to obtain valuable information about the physical properties of both the solar wind and the discrete radio sources. Developing this idea, Ozernoi and Shishov (1980) showed that interstellar scintillation observations could be used to study fine structure (scales of the order of 10^{-5} arcseconds) in quasars and galactic nuclei at centimetre wavelengths. Scintillation observations can appreciably supplement results obtained with Very Long Baseline Interferometery.

1.5 Radio Astronomy Space Studies in the Spectroscopy Laboratory

The first suggestion to place radiometers operating at centimetre and millimetre wavelengths on board spacecraft for radio astronomy studies was put forth jointly by M. A. Kolosov, A. E. Basharinov (Institute of Radio Physics and Electronics) and Salomonovich (FIAN) as early as 1962. The basis for such an experiment was laid and the required equipment developed and tested in collaboration with the groups of S. T. Egorov and L. D. Bakhrakh over the following years. The successful results obtained using a multi-channel radiometer (devised later) to study the surfaces and atmospheres of Mars and the Earth are now well known (see, for example, Essay 2).

Fig. 1.20 "Obzor" submillimetre cryogenic radiometer of FIAN. It was used to obtain maps of the Earth as a planet at wavelengths of 0.1 and 0.5 mm from the "Cosmos-669" spacecraft during five days in 1974

Beginning in 1965, a new direction was developed at FIAN, associated with the expansion of studies at still shorter wavelengths, in the submillimetre range (1–0.1 mm). To exclude the influence of the Earth's atmosphere, which severely hindered the passage of this radiation, it was necessary to develop a submillimetre receiver suitable for use on board balloons, satellites and orbiting piloted stations. This problem was tackled in the Spectroscopy Laboratory, which had experience with space studies, under the supervision of Corresponding Member of the Academy of Sciences S. L. Mandel'shtam. It was worked on by the Group, and then the Section, of Space Submillimetre Studies headed by Salomonovich.

Spectral radiometers and telescopes with liquid-helium-cooled photoresistor-type receivers designed for observations at submillimetre wavelengths were also developed and tested under natural conditions. These were tested on high-flying balloons, *Vertikal'* geophysical rockets, automated satellites and, finally, on board the *Salyut-8* orbiting piloted station. The specialists in infrared studies V. I. Lapshin and A. S. Khaikin (1937–1977), the radio physicists and radio engineers S. V.

Fig. 1.21 Laboratory testing of the BST-1M submillimetre telescope of FIAN, with its parabolic surface and receivers cooled to about 4.5 K. In 1977–1982, several expedition teams carried out experiments with the BST-1M on the "Salyut-6" orbiting manned station

Solomonov, V. S. Kovalev, V. N. Bakun and G. B. Semin, and the specialist in low-temperature physics T. M. Sidyakina and others all actively participated in this work. The setting up of the experimental base for the section was facilitated by P. D. Kalachev.

Original interference filters, Fourier interferometers (Lapshin) and other equipment were developed in 1965–1983. The construction of instruments was carried out by B. N. Leonov and V. N. Gusev. Since the receivers required cooling to low temperatures, the Cryogenics Department of FIAN (A. B. Fradkov, V. F. Troitskii) created unique on-board cryostats for telescopes operating at submillimetre wavelengths under weightless conditions. The first systematic data on the radiation of the Earth as a planet were obtained in this period at 0.1 and 0.3–0.8 mm, which were of interest for studies of the atmosphere. The *Obzor* radiometer (Fig. 1.20) on board the *Kosmos-669* spacecraft was used to make the first maps of the Earth at these wavelengths, which established the relationship between the distribution of the submillimetre brightness temperature and the dynamics of active regions in the troposphere and atmosphere of the Earth. The long (from 1977 through 1981) use of the 1-m BST-1M telescope (Fig. 1.21) on board the *Salyut-6* station made it possible to accumulate valuable information on the functioning of large-scale radio astronomy telescopes on piloted orbiting stations. The work of FIAN on the creation of space radiometers and telescopes for submillimetre observations and studies of the submillimetre radiation of the Earth was highly valued. At the *Vystavka Dostizhenii Narodnogo Khozyaistva* ("Exhibition of National Economic Achievement") of the USSR, the construction of the *Obzor* radiometer was awarded medals (Khaikin, Solomonov, Troitskii and others) and the BST-1M telescope a diploma and medals (Bakun, Salomonovich, Solomonov, Semin and others).

Chapter 2
Radio Astronomy Studies in Gorkii

A.G. Kislyakov, V.A. Razin, V.S. Troitskii, and N.M. Tseitlin

Abstract Stages in the development and main results of studies of the radio emission of the Sun, Moon, planets, Galactic and extragalactic sources at Gorkii are described. Major achievements are noted, including the discovery of a hot core in the depths of the Moon, the polarised Galactic radio background, etc.

2.1 The First Stage (1946–1957)

The development of radio astronomy studies in Gorkii began at Gorkii State University and the affiliated Physical–Technical Institute (GIFTI) in 1946 under the supervision of Professor M. T. Grekhova, who attracted the Department of Electrodynamics and the associated scientific section of GIFTI to this work.

A large influence on the successful development of these studies was exerted by the organisation at Gorkii State University in 1945 at the initiative of A. A. Andronov, Grekhova and G. S. Gorelik of the first Radio Physics Department in the country, whose Dean was Grekhova herself. The students and lecturers in this new Department actively took part in various radio astronomy investigations. In particular, in 1946, I. L. Bershtein began the development of a radiometer designed to operate at 10 cm. This radiometer, which had a threshold sensitivity of about 3 K, was constructed in the middle of 1948, and was subsequently used to measure the radio emission of the Sun, as well as in experiments on the application of radio astronomy equipments and methods to applied problems (M. M. Kobrin, Bershtein). However, this radiometer was not used further for systematic observations, since a more modern device enabling the same measurements under field conditions was devised.

At the beginning of 1947, radio astronomy studies began to be carried out at GIFTI and the General Physics Department of Gorkii State University under the supervision of Gorelik. Work in radio astronomy was initially (1947–1952) coordinated by the Lebedev Physical Institute (FIAN) under the supervision of S. E. Khaikin. A plan defining the specific themes for radio astronomy work at GIFTI was adopted (with the scientific supervision of this work being given by Grekhova, and then Gorelik, starting from 1948).

In 1947, V. S. Troitskii, who was then a PhD student at Gorkii State University (supervised by Gorelik), chose radio astronomy as the theme for his dissertation. He began to work on a radiometer operating at metre wavelengths, which was then considered the most promising wavelength range for radio astronomy observations. The signal in the radiometer was modulated by switching the receiver input from the antenna to an equivalent noise load using a polarised relay at a frequency of about 25 Hz. After detection, the signal was separated out using a narrow-band tuning-fork filter. The radiometer was intended for operation at a wavelength of 4 m together with a receiver and antenna of the student radar detection station. At that time, no modulation-type radiometers for metre-wavelength observations had been developed anywhere. In March 1948, this radiometer was used to carry out the first observations of the solar radio emission. The "waveguide-type" antenna was able to rotate in azimuth, and was located on the roof of a building of the Radio Physics Department in the very centre of town, with the radiometer located in the laboratory. Since the antenna could rotate only in azimuth, observations were conducted either at sunrise or sunset.

In spite of the low sensitivity of the radio telescope and the imperfect operation of the radiometer, they represented a first important step that made it possible to contemplate the solution of theoretical and practical problems in radio physics associated with methods for the reception of weak, noisy signals in general, and at metre wavelengths in particular. In contrast to the situation at centimetre wavelengths, where a Dicke radiometer had already been devised, there were no previous analogs to the technical problem of designing a radiometer for metre wavelengths.

The development of a more modern radio telescope suitable for operation under field conditions at a wavelength of 1.5 m began in the middle of 1948. This wavelength was chosen because two beds of co-phased antennas on a turning mount were available from an old radar station. At the suggestion of Gorelik and Grekhova, two new beds of appreciably larger co-phased antennas that could be installed on the same turning mount were constructed under the supervision of A. P. Skibarko. The development of a laboratory model for the radiometer proceeded in parallel with the manufacture of these antennas. New ideas for signal modulation were tested, and parts for the radiometer were assembled. The student from the Radio Department V. A. Zverev (now a Corresponding Member of the USSR Academy of Sciences) actively took part in this work, proposing a number of signal-modulation schemes and ways to modernise the radiometer design.

When the main principles behind the operation of the new radiometer had been tested under laboratory conditions, the engineers V. L. Rakhlin and A. A. Varypaev and the mechanical technician E. A. Lyubimkov were given the task of making an operational version. In this way, the first experimental radio astronomy group was formed in Gorkii.

The new radio telescope operating at 1.5 m wavelength differed from previous telescopes in a number of interesting features of its design. It incorporated two types of modulation: switching the receiver input from the antenna to an equivalent noise load, and nodding the direction of the antenna beam. As is indicated above, the antenna consisted of two co-phased beds of antennas supported on a common turning

Fig. 2.1 Radio telescope constructed to operate at 1.5 m (Zimenki station)

mount (Fig. 2.1). A high-resistance regulatable segment of terminated co-axial cable in a trombone-like shape taken from the radar equipment was used as an equivalent noise load. This "trombone" made it possible to change the length of the terminated cable from a quarter to half a wavelength, thereby changing the impedance of the input amplifier, and thus the noise level of the load. By setting the noise level in the equivalent load equal to the mean noise level of the antenna, which was determined by the distribution of radio emission received from the sky and the Earth's surface, it was possible to exclude the background signal. In this way, a quasi-null receiver method was realised.

The nodding of the beam was realised by varying the phase difference fed to the antenna beds by alternately attaching the receiver input to two specially adjusted feeder points (joining both antenna beds) using a relay contact or volume switch operated by a synchronised motor. The nodding angle was chosen to be half the antenna-beam width. The low-frequency part of the radiometer had a heterodyne filter analogous to a Dicke radiometer.

The beam-nodding method proved to be very effective, enabling reliable operation in the presence of undirected interference in Gorkii, where the radio telescope was used to conduct test observations of the Sun in Spring 1949. The telescope was then moved to the Zimenki Station on a high bank of the Volga 30 km from Gorkii, where regular observations of the solar radio emission at 1.5 m began in August 1949. During this same time, measurements of the refraction of metre-wavelength radio waves in the Earth's atmosphere were also carried out using observations made during the radio sunrise and radio sunset.

Various researchers and PhD and other students at the university took part in studies using this first Gorkii radio telescope—S. A. Zhevakin, G. G. Getmantsev, A. I. Malakhov, A. V. Zolotov, V. M. Plechkov, Z. I. Kameneva and others. Starting at this time, Zimenki began to be developed as a radio astronomy laboratory outside the city. A two-hectar plot of land was given to this laboratory in 1949, two prefabricated buildings were put up in 1952, and a laboratory building, apartment building, dormitory and helium cryogenic station were constructed and brought into use in 1960–1962.

The metre-wavelength radio telescope was used right up until 1956. Radio refraction studies were carried out until 1952. Later, Razin used the telescope to search for polarised cosmic radio emission. Using the existing radiometer together with new, specialised radiometers that were better suited for detecting polarised signals, Razin demonstrated the presence of linear polarisation of the extended cosmic radio emission in 1956. Getmantsev used this same radio telescope to measure the radio emission of the most powerful discrete sources in order to accurately determine their intensities. For this purpose, he developed a means for thermal calibration of the radiometer to replace the previously used noise-diode calibration.

The successful operation of the radiometer in the beam-nodding mode stimulated the use of this method with a new radio telescope operating at 10 cm wavelength, also based on a radar station. In this latter case, there was a device to nod the beam using mechanical rotation of the feed (dipole) about the main focus of a paraboloid. In essence, this was a nearly complete radio telescope, which provided an important function of a radiometer—modulation of the signal. It was only required to connect a heterdyne filter. Rakhlin began the development of this instrumentation, completing this work in March 1950.

Systematic measurements of the radio emission of the Sun from sunrise to sunset were carried out on this radio telescope in the Spring and Summer of 1950, in order to determine the refraction and absorption of radio waves in the Earth's atmosphere and study seasonal variations in these phenomena. The radio emission of the Earth's atmosphere was first detected at such a long wavelength (10 cm) using this telescope.

The circular rotation of the beam of the radio telescope enabled very accurate determination of the direction toward the Sun. This was a prototype radio sextant, which subsequently found wide application thanks to theoretical and experimental development of this idea with the participation of Kobrin, I. F. Belov, Rakhlin and B. N. Ivanov.

At the end of 1948 and the beginning of 1949, the development of a radiometer designed to operate at 3 cm was begun at the initiative of Grekhova. This work was carried out in the Electrodynamics Department by the group of S. I. Averkov. The receiver part of an on-board aviation radar station was used. A Dicke scheme with a rotating absorbing disk was used to modulate the signal. In April–June 1950, measurements of the radio emission of the Sun and of the Earth's atmosphere were carried out on the 3-cm radio telescope under a protective cap; the telescope's sensitivity was about 5 K and its reflecting surface had a diameter of 0.6 m.

The development of radio telescopes and methods for measuring weak continuum signals were accompanied by theoretical studies by Troitskii. The creation of the first metre-wavelength radio telescope and accompanying studies formed the basis for Troitskii's PhD dissertation, which was the first dissertation on radio astronomy in the Soviet Union, defended in Spring 1950 at FIAN.

Fig. 2.2 Samuil Aronovich Kaplan (1921–1978)

From the very first, the development of radio astronomy in Gorkii was closely tied to the solution of topical practical problems, first and foremost with studies of the conditions for the propagation of radio waves in the Earth's atmosphere. Radio astronomy presented unique possibilities for studies of the refraction and absorption of radio waves in the atmosphere. These phenomena were first studied based on the attenuation of the radio emission of the Sun, and then also the Moon, as a function of the hour angle of the Sun or Moon during the course of the day. Later, it became possible to determine the atmospheric absorption based on the thermal radio emission of the atmosphere itself, which was reliably measured at centimetre and decimetre wavelengths using the radio telescopes that had been built. The need to develop the theory of radio emission of the atmosphere and methods for measuring absorption arose. Work in these directions was carried out by Troitskii, Zhevakin and others. Later, wide-ranging theoretical, methodological and experimental studies of the absorption of radio waves in the atmosphere and of the radio emission of the atmosphere itself from submillimetre to decimetre wavelengths were conducted by Zhevakin, N. M. Tseitlin, A. G. Kislyakov, Razin, K. S. Stankevich, L. I. Fedoseev, V. V. Khrulev, D. A. Dmitrenko and others.

Thus, in 1946–1950, Gorkii State University, together with the Gorkii Physical–Technical Research Institute and FIAN, became a leading centre for radio astronomy studies in the Soviet Union. At that same time, studies in theoretical radio astronomy began to develop in Gorkii under the supervision of V. L. Ginzburg, with whom Getmantsev, N. G. Denisov, Razin, Zheleznyakov and others collaborated. The well known Soviet astrophysicist S. A. Kaplan (Fig. 2.2) also worked in Gorkii from 1960 until his tragic death in 1978.

2.1.1 *Radio Telescopes and Radiometers*

After the first results of radio astronomy and applied studies, it became evident that further development of radio astronomy would require new, more powerful radio telescopes. This was especially true for work at centimetre and decimetre wavelengths, where the first telescopes used had diameters of less than one metre. The director of the GIFTI Grekhova and the division head Gorelik ordered 4-m-diameter antennas to be built in accordance with plans provided by FIAN (P. D. Kalachev) from one of the Gorkii factories. The antennas were manufactured by 1952, but without a means to turn them. Turning mounts from decommissioned zenith cannons were adopted for this purpose. In this way, this wartime technology was turned toward peaceful, scientific goals.

The development of new, more modern and sensitive radiometers operating at centimetre and decimetre wavelengths was begun in the radio astronomy laboratory that formed in Gorelik's division. Experience working on new radio telescopes operating at 3 and 10 cm showed a number of important problems with the radiometers used, which had a fundamental character. This was especially true of the Dikke-type radiometer working at 3 cm. First and foremost, the large background signal, equal to the difference between the temperatures of the antenna (usually 50 K) and the disk (300 K) hindered measurements of weak signals due to fluctuations in the amplification. In addition, a strong false signal usually arose due to interference between the input noise signals reflected from the antenna and the disk, which were not perfectly matched with the waveguide. There were also cruder technical problems, such as the modulation of the mixer noise at the input, etc.

To eliminate the background signal and expand the capabilities of the radiometer, it was decided to devise a new radiometer designed to operate at 3 cm in which the receiver input would be switched between two equivalent inputs—one connected to the antenna feed and the other to an adjustable noise source or to a second antenna feed located next to the first, in order to implement nodding of the beam, which in fact turned out to be very effective. This required a rapidly acting switch. At that time, there were only spark switches, which were not suitable for this purpose due to their high noise. Therefore, a mechanical, rapidly acting, waveguide T-joint switch operating with a frequency of 20–80 Hz was created. Through the action of a polarised relay, the switch was alternately closed first one, then the other arm of the T-joint. The closure of the circuit was brought about by closing a gap between pins in the clearance of the inductive diaphragms. The gap was closed in antiphase by a rod moved by a relay inside the tube-like pins.

A switch amplifier with two equivalent inputs was also used in the 10-cm radiometer. Another innovation that appreciably improved the real sensitivity of the radiometer was the implementation of a "snail shell"—a long waveguide curled up like a snail's shell that eliminated the interference of the input noise signals of the receiver. When the length of the waveguide used is such that the time for the noise signal to travel to the antenna and back is longer than the noise-correlation time in the receiver bandwidth, there will be no interference. In another method used to suppress the interference signal, the frequency of the heterodyne was modulated over a small range determined by the length of the antenna tract; when the frequency was rapidly changed periodically with a period that was shorter than the time constant of the output, the interference pattern was averaged. All these techniques made it possible to exclude false signals and various "parasitic" effects, enabling the obtained sensitivity to approach its theoretical value. The properties of interference noise were studied by Khrulev and A. M. Starodubtsev, whose theory and the associated influence on the measurement accuracy were considered in one of the papers of Troitskii.

The desire to understand all sources of errors in measurements of the noise signals led further to studies of the noise in the heterdynes used in all radiometer applications at that time. Questions having to do with generator noise had already been studied before the war by I. L. Bershtein, who developed the theory of natural line widths and a method for measuring this width in radio generators. In 1954, Troitskii devised an apparatus for this purpose and measured the natural line widths of a klystron generator operating at 3.2 cm. The results of this work provided a basis for choosing an intermediate frequency for the radiometer receiver and an optimum heterodyning scheme.

Together with such experimental studies of noise in systems with self-excited oscillations, a number of theoretical investigations were carried out. In particular, a more adequate mathematical approach to studying fluctuations of the self-excited oscillations was developed in Gorkii and Moscow (S. M. Rytov). Instead of applying the Einstein–Fokker equations to solve the fluctuation equation, a more usual spectral approach was used. This enabled Troitskii to resolve questions having to do with the influence of flicker noise and taking into account periodic non-stationarity of interacting shot noise signals.

When striving to make accurate measurements of noise signals, the sensitivity threshold was not as important as excluding systematic errors in the measurements. This required accurate calibration, which could be provided only by standard sources of thermal radio emission. Therefore, as part of the development of the new radiometers (for 3 and 10 cm), precise standard sources of thermal radio emission with fairly high temperatures of 200°C or higher were devised. Such a standard source was made by Rakhlin for the waveguide used at 3 cm, based on a matched blackbody wedge that was heated in the waveguide. This long served as the primary standard for calibrating radiometers at this wavelength. Simultaneously, Plechkov developed a standard source for a temperature of 1500°C in the form of a tungsten-filament lamp with a special construction, which was placed in the waveguide. This was used as a secondary standard, and required calibration using a primary standard source.

At the same time, primarily through the efforts of students, work was begun on a radiometer designed to operate at 1.63 cm wavelength, which was brought into use by Kislyakov and L. I. Turabovaya in 1954, and was used for observations of a solar eclipse in Novomoskovsk (Fig. 2.3).

During the course of this work on various radiometers, the idea arose of using them as devices to accurately measure the power of weak signals and the noise signals of various instruments, and also to calibrate attenuators. The first industrial radiometers operating at 3 and 10 cm were devised, based on studies and developmental work carried out by Rakhlin and Starodubtsev in 1954–1960 under the supervision of N. A. Serebrov. In subsequent years, serially produced, broadband radiometers adjustable to wavelengths from 2 to 60 cm were developed at this same institute, under the supervision of N. N. Kholodilov.

Fig. 2.3 Centimetre-wavelength (1.6, 3, and 10 cm) radio telescopes at the solar-eclipse expedition of 1954 at Novomoskovsk

2.1.2 Radio Astronomy Observations

Intensive radio astronomy studies began in 1952. Razin performed the first measurements in the world of the radio emission of Cassiopeia A and the Crab Nebula at 3 and 10 cm using the new 4-m parabolic radio telescopes. In 1954, these measurements were again carried out by Razin and Plechkov, and Razin who began investigations aimed at searching for polarisation of the extended cosmic radio emission. With the development of the theory of the synchrotron radiation of the Galaxy, it because obvious that the extended cosmic radio emission might be significantly polarised. However, the degree of polarisation of the radio emission received at the Earth should be small (a few percent or less) due to the inhomogeneity of the magnetic fields in the interstellar medium. Razin first pointed out the importance when attempting to detect this polarisation of taking into account the rotation of the plane of polarisation due to Faraday effects in the interstellar medium and the ionosphere of the Earth. He proposed a means to use this effect, consisting of modulating the bandwidth of a radiometer in order to reliably distinguish the polarised component of the radiation. This drastic improvement in the radiometer enabled the detection of linear polarisation of the extended cosmic radio emission at 1.45 m in Summer 1956. On May 10, 1964, Razin was awarded a diploma for this discovery, with the discovery date indicated as June 1956.

In the beginning of the 1950s, a large number of studies of the solar radio emission during eclipses were conducted. Observing during eclipse made it possible to observe local sources on the Sun with resolutions of arcminutes. In addition, such observations provided information about the radio diameter of the Sun. The data obtained at various wavelengths were used to construct a model for the solar corona, that is, the dependence of the kinetic temperature and electron density on height above the solar photosphere. It was also of interest to measure the radio brightness distribution across the solar disk. These questions were investigated in a number of expeditions. The first such expedition using Gorkii radio telescopes was in 1952 to the settlement of Archman, near Ashkhabad, which fell within the total-eclipse strip. Radio telescopes operating at 3 and 10 cm wavelength were used.

In 1954, there was another expedition to the Novomoskovsk region, where likewise, it was possible to observe during a total solar eclipse. Three radio telescopes, operating at 1.6, 3 and 10 cm were used.

Finally, a Radio Physical Institute expedition observed a solar eclipse in China in 1958, on the Hainan Island, using radio telescopes operating at 1.6, 3 and 10 cm. A. N. Malakhov, Razin, Rakhlin, K. S. Stankevich, K. M. Strezhneva, Khrulev, N. M. Tseitling and other participated in this expedition, which was supervised by Troitskii.

Careful measurements of the absorption of radio radiation in the Earth's atmosphere from its own radio radiation were conducted in 1952, in various seasons. Numerous data on the total absorption of radio radiation for various absolute humidities at the Earth's surface were obtained. Gorelik pointed out to Tseitlin, who was conducting these studies, that the dependence of the absorption on the absolute humidity seemed to be linear. In this way, a method was devised to separately determine the absorption by oxygen (extrapolating the linear relation between the total absorption and the humidity to zero humidity) and by water vapour (from the slope of the linear relation). This method led to the beginning of further investigations of absorption in the atmosphere over a wide range of wavelengths, from millimetres to several tens of centimetres.

In 1952–1955, radio astronomy methods for measuring the parameters of antennas began to be developed. As a rule, the antenna beam was usually measured using the radio emission of the Sun. It was at that time that Razin and Troitskii proposed and developed a method for measuring the losses in the antenna tract and in the antenna itself using measurements of the antenna noise.

Radio astronomy methods for studying antennas subsequently developed in the independent direction of applied radio astronomy due primarily to work by radio astronomers and radio physicists in Gorkii (Troitskii, Tseitlin, Dmitrenko and others).

Investigations of the radio emission of the Moon were begun in 1951–1952 under the supervision of Troitskii. The new 4-m radio telescopes enabled the reliable detection of the thermal radio emission of the Moon. By that time, Piddington and Minnet in Australia had detected a phase dependence for the radio emission at 1.25 cm, and established that the amplitude of the Moon's temperature fluctuations was much lower than at infrared wavelengths. They explained this by suggesting that the radiation at 1.25 cm was not emitted at the surface, where the temperature fluctuations from the lunar day to night were 300 K, but instead from some depth where these fluctuations were damped by a factor of five.

The first observations of the Moon were carried out at 3 cm, but only when the Moon was rising or setting, in order to determine the refraction and absorption of the radio emission in the atmosphere. These measurements were used to estimate an upper limit for possible temperature fluctuations of the Moon, which turned out to be much lower than at 1.25 cm. It became obvious that this wavelength was emitted from a still thicker layer of material, at a depth where the temperature fluctuations were essentially completely damped. Troitskii studied the properties of the surface-layer material and devised a very complete theory for the lunar radio emission for the case of a homogeneous composition of its surface layer. Specialised, more accurate studies of the lunar radio emission were then begun. Much attention was paid to accurate calibration of the fluxes measured by the radio telescope. The intensity of the solar radio emission was measured using a reference horn antenna, and this intensity was then used to calibrate the 4-m radio telescope.

These studies led in 1955 to the detection and measurement of the phase behaviour of the lunar radio emission at 3 cm. These measurements, in turn, represented the beginning of a more than ten-year series of investigations of the Moon from millimetre to decimetre wavelengths, which served as the basis for the development of new methods for accurate absolute intensity measurements and the calibration of radio telescopes by Troitskii and Tseitlin, and later V. D. Krotikov and V. A. Porfir'ev as well. Kislyakov began the work on the creation of radiometers for millimetre wavelengths. It was necessary to overcome enormous difficulties, since neither measuring devices nor the necessary microwave elements (in particular, crystal mixers) existed for this wavelength range.

In 1954, work on radar studies of the Moon began at Gorkii, whose importance had been pointed out earlier by Academician N. D. Papaleksi. Radar measurements of the Moon at centimetre wavelengths (3 and 10 cm) were first realised in 1954–1957 by a group of researchers under the supervision of Kobrin. This work encountered great difficulties, since, in contrast to the situation at metre and decimetre wavelengths, there were not yet at that time transmitters operating at centimetre wavelengths that could provide the necessary radio power for this purpose. Researchers at the Physical–Technical Institute had to carry out special studies to learn how to design and construct the necessary transmitting and receiving equipment. They developed the method of bistatic radar measurements using comparatively low power and a modest antenna area, foreseeing the use of radiometer devices to receive the reflected signal, and proposed a phase method for the measurement of the distance to the Moon.

To obtain high power, a method for adding the powers of several (two to four) centimetre-wavelength generators in series using bridge connectors was proposed and realised. These systems were able to yield continuum radio powers of several kilowatts and a frequency stability that was higher than that of the individual generators making up the system (D. I. Grigorash, V. S. Ergakov). At the same time, methods for the parametric stabilisation of the transmitter and monitoring of this stabilisation using a quartz heterdyne, as well as methods for amplitude manipulation, were developed in parallel. Specialised radiometer receivers with a travelling wave tube at the input and a narrow receiver bandwidth of from 1.5 to 3.0 MHz were devised to increase the sensitivity of the receivers (V. I. Anikin, I. M. Puzyrev). The heterodyne of the 3-cm receiver was stabilised with a quartz self-tuning phase system. I. V. Mosalov and V. I. Morozov designed and built antenna systems with accurate 4–5-m diameter surfaces and semi-automated tracking (an equatorial mount).

The transmitter was in an institute building in Gorkii, while the receiver was located 30 km away at the Zimenki site. The frequencies of the transmitters and receiver heterodynes were matched using a special communication channel. The transmitter and receiver were turned on alternately for 3 s in order to exclude the direct reception of the transmitted signal.

During the lunar-radar experiments at 10 cm in July 1954 and at 3 cm in June 1957, it was found that the effective area of the Moon for reverse scattering[1] was 0.07 of the area corresponding to its transverse cross section, in agreement with later measurements at centimetre wavelengths obtained outside the Soviet Union. It was found that the Moon's reflection of centimetre waves was very different from its reflection of optical waves. Based on the dependence of the signal on the accuracy with which the antenna was pointed at the centre of the Moon, it was concluded that the reflection occurs primarily from the central part of the visible lunar disk.

In this period and also in subsequent years, theoretical studies of radio astronomy and astrophysics were also widely developed in Gorkii. Important results in the theory of the extended cosmic radio emission and the radio emission of discrete sources were obtained by Ginzburg, Getmantsev and Razin, in studies of plasma mechanisms for the generation of various types of solar radio emission by Zheleznyakov, and in connection with various problems in astrophysics by Kaplan.

At that same time, Ginzburg and Getmantsev developed the theory of occultation observations of the Crab Nebula, when the Moon passes in front of the source for some period of time. This method, which was also put forth by Gorelik, is still used today, and provides an effective resolution of fractions of an arcsecond.

2.2 Development of Radio Astronomy at the Radio Physical Research Institute

The number of radio astronomy studies being undertaken was appreciably expanded with the formation in 1956 of the Radio Physical Research Institute (NIRFI), on the basis of several departments of the Physical–Technical Institute and research groups at the Polytechnic Institute.

Three radio astronomy departments were created in the institute: the Department of Microwave Radio Astronomy, headed by Troitskii, the Department of Long-wavelength Radio Astronomy and Ionospheric Physics, headed by Germantsev and the Department of Solar Radio Emission, headed by Kobrin. In subsequent years, three new divisions were created from the Department of Microwave Radio Astronomy: the Laboratory of Galactic and Extragalactic Radio Astronomy in 1963 (headed by Razin), which became a department in 1967, the Department of Applied Radio Astronomy (headed by Tseitlin) and the Laboratory of Millimeter-wave Radio Astronomy (headed by Kislyakov), which became a department in 1975. A new

---

[1] The ratio of the power of an isotropic source of radiation located at the same place as the observed object that would give rise at the receiver to the same power flux density as the observed object, to the power flux density of the incident radiation from the object.

department headed by E. A. Benediktov became distinct from the department of Germantsev in 1965, and another new department headed by V. O. Rapoport was separated off in 1976. In addition to studies of radio-wave propagation and plasma physics, these departments are concerned with the decametre radio emission of the Sun and powerful discrete radio sources.

The radio astronomy instruments at the Zimenki station (which became the Zimenki Laboratory in 1971) were expanded and outfitted. New radio astronomy stations of NIRFI were founded in Karadag (Crimea) and Staraya Pustyn' (Gorkii region) in 1964, and a radio astronomy base near Vasil'sursk (Gorkii region) in 1965, which later became the Vasil'sursk Laboratory.

Systematic radio astronomy studies in a number of directions began in the newly founded institute: investigations of the radio emission of the Sun, Moon and atmosphere, the polarisation of cosmic radio emission and the intensities and spectra of discrete radio sources, as well as work in applied radio astronomy. These studies were carried out on the basis of corresponding theoretical, methodological, instrumental and antenna developments.

Close and fruitful collaboration with radio astronomers at FIAN under the supervision of S. E. Khaikin and then V. V. Vitkevich was important for the development of radio astronomy studies at NIRFI.

2.2.1 The Zimenki Laboratory

A fully steerable 5-m millimetre-wavelength radio telescope was installed at Zimenki in 1957 (Kalachev, Mosalov, Morozov and others). In 1958, work began on the creation of instrument complexes for studies of cosmic radio emission at decametre wavelengths and studies of the ionosphere using radio astronomy techniques (Benediktov, V. V. Velikovich, Getmantsev, L. M. Erukhimov, Yu. S. Korobkov, A. I. Tarasov and others), the construction of two 15-m fully steerable radio telescopes, which were completed in 1962 and a complex of laboratory buildings and living quarters (Kobrin, A. A. Petrovskii, Zhuravlev, L. V. Grishkevich, I. S. Motin and others; Fig. 2.4).

In 1967, the unique RT-25 millimetre-wave radio telescope was constructed and brought into use. This is a transit-type telescope with a 2×25 m^2 surface, and has until recently provided the highest angular resolution at millimetre wavelengths in the world—$13''$ in azimuth and approximately $2'$ in elevation at 1.35 mm (Kislyakov, Mosalov, V. P. Gorbachev, V. N. Glazman, K. M. Kornev, V. I. Chernyshev and others).

In 1976, a "Solar Service" complex was established at Zimenki, consisting of centimetre- and decimetre-wavelength radio telescopes and standard "black" disks, which provide high-accuracy monitoring observations of the solar radio flux (O. I. Yudin, M. S. Durasova, T. S. Podstrigach, Yu. B. Bedeneev, I. M. Prytkov, G. I. Lupekhin, G. A. Lavrinov, V. S. Petrukhin and others).

2 Radio Astronomy Studies in Gorkii

Fig. 2.4 15-m radio telescope, the system of smaller radio telescopes, and the black disk for the solar service at Zimenki

2.2.2 The Karadag Radio Astronomy Station

The Karadag Radio Astronomy Station of NIRFI was established as a base for making very accurate absolute measurements of the intensity of the radio emission of the Moon and of powerful discrete sources at centimetre wavelengths. In 1957, Troitskii and Tseitlin proposed the method of comparing the received radio emission from a source with the radio emission of a standard black area. Such an area can be realised by observing the radiation from a hill slope below the Brewster angle. A metallic disk that reflects the "cool" radio emission from the region of the zenith toward the antenna is also placed on the hill side. The difference of the intensities of the "black" hill side and the metallic disk served as the calibration signal.

The landscape along the shore of the Black Sea between Planerskii and Sudak, where measurements were carried out under expedition conditions in 1957–1959, turned out to be a convenient location to realise this method (Troitskii, Tseitlin, Porfir'ev, Rakhlin and others). However, these measurements did not provide sufficient accuracy due to diffractional "heating" of the metallic sheet (radio emission from the Earth fell into the diffractional beam of the sheet and was reflected toward the antenna). Therefore, it was decided in 1960 to use black disks observed at fairly large angles to the horizon as standards, in order to shield them from "cool" regions of atmospheric radio emission. The first experiment of this kind was carried out in 1960 at Yalta, where a half-metre disk was mounted on the roof the *Oreanda* hotel, and radio telescopes operating at wavelengths of 3 cm (with a diameter of 1.5 m) and 10 cm placed 200 m from this location. Since the angular diameter of the disk was equal to the angular diameter of the Moon (which facilitated more accurate measurements of the lunar radio flux), it was called an "artificial Moon." To take into account radio emission from the Earth (the so-called diffraction correction), the radio emission of a screen with an opening equal in size to the disk was used in addition to the radio emission of the standard black disk itself (Troitskii, Porfir'ev, Krotikov). After the corresponding calculations of the diffraction correction carried out by Tseitlin, it was possible to avoid using the screen with an opening, and

Fig. 2.5 Five-metre artificial Moon at the peak of a cliff (Karadag station, Crimea)

measurements began to be conducted by comparing the source radio emission and the radio emission from the standard black disk on its own. In this way, the "black disk" (or "artificial Moon") method arose—one of the most accurate methods for obtaining absolute measurements of the intensities of radio signals.

Since the placement of the black disk required a high pole or hill, for economic reasons, it was most convenient to conduct accurate measurements of the intensities of the Moon and discrete radio sources at sites in the mountains. After extensive study, it was determined that the most suitable mountainous site to carry out measurements was a region in the Eastern part of the Crimea. A large expedition to Sudak took place in 1961 (Troitskii, Tseitlin, Stankevich, Krotikov, Porfir'ev, L. N. Bondar', K. M. Strezhneva and others), where an artificial Moon with a diameter of 4 m was place on a cliff near Genuez Fortress (Fig. 2.5). The radio telescopes were placed on the sea shore at a distance of about 400 m from the disk. The main task of the expedition was to study the Moon at 3 and 10 cm with the aim of determining a possible temperature gradient with depth.

As a result of this work, the increase in the temperature of the Moon with depth was reliably established for the first time—a discovery for which Troitskii and Krotikov were awarded a diploma with the discovery date indicated as November 19, 1962.

At the beginning of 1961, Razin and Tseitlin proposed to carry out precision measurements of the fluxes of the most powerful discrete sources using the "black" disk method, at decimetre wavelengths as well as 3 cm. In other words, they wished to compile an accurate catalogue of the absolute fluxes of these sources over a wide range of wavelengths.

To realise this programme and programmes of studies of the lunar radio emission, a second expedition to Sudak was organised in 1962. Telescopes operating at centimetre and decimetre wavelengths were used. However, the use of larger radio telescopes required that the distance to the black disk be increased, which was not possible at Sudak.

Therefore, decimetre-wavelength measurements of the radio fluxes of the Moon and discrete sources were carried out in 1962–1964 in the Karadag valley. However, here as well, the possibilities for increasing the calibration distances were limited, leading to a search for a "convenient" mountain and surrounding area where it would be possible to separate the black disk and radio telescope by up to 1 km, as was required by the use of large antennas. (The distance from the antenna to the disk must be more than $2D^2/\lambda$, where D is the antenna diameter and λ is the wavelength.) In addition, the mountain must be high enough for the angle at which the artificial Moon was observed to be as large as possible. A suitable place was chosen by Porfir'ev in a clearing on the Karadag massif. The cliff looked so inaccessible that it appeared it might be impossible to mount the disk on it. However, systematic decimetre-wavelength measurements of the fluxes of the Moon and powerful discrete sources began in Autumn 1965. A new artificial Moon was constructed and mounted.

In 1965–1966, radio telescopes with diameters of 4 and 12 m were brought into use, an electrical transmission cable laid, and wooden laboratory buildings and living quarters constructed. In this way, the Karadag radio astronomy station of NIRFI arose, and has been continually operational since.

In subsequent years, another radio telescope with an accurate 7-m surface was installed, and a multi-frequency metre-wavelength interferometer and stand for studies of the sporadic radio emission of the Earth's magnetosphere constructed at the station, which is supervised by Stankevich.

2.2.3 The Staraya Pustyn' Radio Astronomy Station

The Staraya Pustyn' radio astronomy station of NIRFI was founded by Razin in 1964 for studies of the polarisation and spectrum of the Galactic radio emission.

The first polarisation observations there were obtained in 1965 with an 8-m fully steerable radio telescope (Razin, A. N. Rodionov, Khrulev and others). A 12-m telescope was brought into operation in 1967 (later increased to a diameter of 14 m in 1975), a 7 m telescope in 1970–1971 and two 14-m decimetre- and metre-wavelength telescopes in 1980–1981. Currently, the station has three 14-m radio telescopes that form a polarisation-sensitive interferometer with variable baselines (Fig. 2.6). Both fundamental and applied research was carried out using these telescopes (Razin, A. I. Teplykh, E. N. Vinyaikin, L. V. Popova, Khrulev).

In 1967–1971, two fully steerable 7-m radio telescopes operating at centimetre and decimetre wavelengths were brought into use, together with two 25-m towers with standard black disks separated from the antennas by 100 m and 50 m (Zhuravlev, E. A. Miller, Mosalov and others; Fig. 2.7). These radio telescopes were used to realise the method proposed by Tseitlin of obtaining accurate absolute measurements of radio fluxes by focusing the antennas (during calibration observations)

Fig. 2.6 Three-antenna polarisation radio-interferometer with a variable baseline (Staraya Pustyn' station)

Fig. 2.7 Seven-metre radio telescope and tower with the black disk (Staraya Pustyn' station)

on a disk located in the Fresnel zone. New radio astronomy and amplitude–phase (radio holographic) methods for studying antennas were also tested on these instruments (Tseitlin, Dmitrenko, A. A. Romanychev, V. I. Turchin, Yu. I. Belov, N. A. Dugin, A. L. Fogel', V. S. Korotkov and others).

In 1978, the RTVS-12 12-m decimetre-wavelength frame-and-guy radio telescope was constructed (Mosalov, N. V. Bakharev, Dugin and others), based on prestressed constructions for antenna surfaces. In 1980–1981, a two-element radio interferometer consisting of two 7-m radio telescopes separated by 417 m East–West operating at decimetre wavelengths as an Earth-rotation aperture synthesis system

Fig. 2.8 12-m radio telescope and five-metre black disk (Staraya Pustyn' station)

Fig. 2.9 German Grigor'evich Getmantsev (1926–1980)

was brought into use, together with a 5-m black disk intended for use with the RTVS-12 telescope for absolute measurements of the intensities of discrete sources and of the extended radio emission at decimetre and metre wavelengths (Fig. 2.8).

2.2.4 The Vasil'sursk Laboratory

The Vasil'sursk Laboratory is located in the Gorkii region near the place where the Volga and Sura rivers merge. It was organised in 1965.

From the first stages of establishing the laboratory, at the head of all undertakings was Getmantsev, whose untimely death occurred in 1980 (Fig. 2.9). Under his supervision, a well equipped base for scientific studies of non-linear phenomena in the ionosphere, radio astronomy and the propagation of radio waves in the ionosphere was constructed, which formed the basis for the Sura experimental complex built in 1980.

In 1968–1969, a metre-wavelength radio telescope with a 30×50 m antenna (rotating in azimuth) whose axis was oriented at $19°$ to the horizontal was brought into use. A fully steerable 8-m decimetre-wavelength radio telescope was also constructed.

2.3 Main Directions for Radio Astronomy Research at NIRFI

2.3.1 Radio Astronomy Studies of the Moon

Experimental studies of the lunar radio emission began at the Zimenki site in 1950 (Troitskii, Kislyakov, Krotikov, Starodubtsev, V. N. Nikonov, Tseitlin, Fedoseev, A. I. Naumov and others). Over the following two decades, a large series of experimental and theoretical studies were carried out, as a result of which:

– data on the temperature regime, composition and structure of the upper surface layer of the Moon to a depth of 10 m were obtained;
– the thermal and electrical characteristics of the lunar soil were determined;
– the growth of the temperature with depth was detected, and the existence of a hot core in the Moon demonstrated.

A model for the lunar soil was proposed based on these studies, which was used in the planning of the landing modules for the *Luna* series of automated interplanetary stations and the body of the lunar walker. The results of the ground-based radio astronomy observations were confirmed by direct measurements on the lunar surface and laboratory studies of the physical characteristics of the lunar soil.

2.3.2 Radio Astronomy Studies of the Sun

The first solar radiospectrographs, covering frequencies from 1 to 12 GHz with a frequency resolution of 60–100 MHz, were devised at NIRFI (Kobrin, Yudin, Podstrigach, Durasova, Vedeneev, A. I. Korshunov, V. M. Fridman, Prytkov, Rapoport, V. V. Pakhomov and others). A series of observational studies of the structure of the radio spectra of local sources and flares was carried out on the 22-m radio telescopes of FIAN and the Crimean Astrophysical Observatory.

Thanks to the high frequency resolution of the spectrographs, fine spectral features were detected in the slowly varying and flare components of the solar microwave emission. Theoretical models for active regions suggested that one possible origin for this fine structure was the presence of inhomogeneities, in particular, neutral current sheets, which are now widely invoked to explain the development of solar flares.

The investigations that have been conducted have opened new paths for studies of active regions on the Sun (in particular, searches and diagnostics for neutral current sheets), as well as pre-flare and flare processes.

On the basis of the unique T-shaped UTR-2 decametre-wavelength antenna near Kharkov and due to the efforts of both NIRFI and the Institute of Radio Physics and Electronics, a solar radio astronomy complex (radiospectrograph, two-dimensional heliograph and radiometer) was constructed and brought into operation, enabling the recording of dynamical spectra of peculiar decametre radio flares and the determination of the locations and sizes of their sources in the solar corona with high resolution.

The UTR-2-based radio astronomy complex is used for investigations of the main properties of the sporadic solar emission at decametre wavelengths, at frequencies of 10–26 MHz. Important observational data on the harmonic modes of type-III flares, the harmonic structure of double III6+III events and the dynamics of the electron flows exciting these flares have been obtained. New and extremely interesting phenomena have been discovered, such as radio echoes of short, narrow-band flares. These studies have appreciably increased our understanding of the physical conditions in the upper corona and the mechanisms generating the sporadic solar radio emission. The detection and investigation of quasi-periodic fluctuations of the solar radio emission stimulated similar studies at other observatories, in particular, in Cuba and the German Democratic Republic.

Beginning in 1966, systematic monitoring observations of the solar radio emission at 9100, 2950, 950, 650, 200 and 100 MHz were conducted at the Zimenki Laboratory. These observations enabled studies of the relationship between the brightnesses of solar flares and radio flares at centimetre wavelengths. They established the temporal connection between radio flares and soft X-ray emission, and the empirical relation between the intensity of the fluxes of protons with energies of 5, 10, 30 and 60 MeV and the integrated fluxes from radio flares at 3 and 10 cm. A statistical relationship between the maximum intensity of the flux of high-energy protons generated in solar flares and the time for the bulk of the protons to travel to the Earth was also detected. This relation can be used to achieve more accurate prognoses of proton fluxes.

2.3.3 *Studies of the Radio Emission of Discrete Sources*

Regular absolute radio astronomy measurements of the flux densities and spectra of powerful discrete sources (Cassiopeia A, Cygnus A, Taurus A, Virgo A) and of the planet Jupiter have been conducted for over 20 years at the Karadag and Staraya Pustyn' sites (Troitskii, Razin, Tseitlin, Stankevich and others). These studies make use of small (diameters of 7–14 m) radio telescopes calibrated using the radio emission of black disks mounted in both the Fraunhofer zone of the antenna (Karadag) and at closer distances, in the Fresnel zone (Staraya Pustyn'). These measurements have yielded precision (with uncertainties of a few percent) spectra of these powerful discrete sources, which have been used throughout the world as standards. These first reference measurements form the basis for a unified (Southern and Northern hemispheres) absolute scale for the spectra of secondary and tertiary calibration sources. The spectral energy distributions of 11 sources from the Cambridge and Parkes catalogues having fluxes of 10^{-25} W/(m^2 Hz) at a frequency of 1 GHz were measured on large telescopes calibrated using the primary standards. These sources form a group of secondary standards, which can be used to determine the parameters of antennas with diameters of 60–70 m. The NIRFI system of standards was used to measure the parameters of the unique 70-m radio telescope with its quasi-parabolic surface.

Investigations of the evolution of young radio supernova remnants have been carried out over some 20 years. Methods for absolute and relative radio astronomy measurements developed at NIRFI were applied in order to accurately determine variations in the frequency distribution of the intensity; a number of studies were carried out using antennas with high resolutions. These observations enabled the detection of secular decreases (fractions of a percent per year) in the radio flux densities of the Crab, Tycho Brahe and Kepler supernova remnants.

In the case of the young supernova remnant Cassiopeia A, the NIRFI measurements revealed a dependence of the rate of the flux decrease on the frequency, time variations of the radio spectral indices, strong time variations in the rate of the flux decrease and intense lines within a limited range of frequencies in the metre-wavelength radio spectrum. Variability in the flux density of the Crab supernova (Taurus A) reaching 15–20% and with a period of the order of several years was also discovered.

Lunar occultation observations (1964 and 1974) were used to study the structure of the Crab Nebula. Variations in the radio brightness distribution and a shift of the centre of gravity of the radio brightness were detected, as well as a relationship between these phenomena and the activity of the central region of the nebula containing the pulsar. The RATAN-600 radio telescope and a feed with linear-polarisation switching developed at NIRFI were used to obtain one-dimensional distributions of the linear polarisation of the radio emission of the Crab Nebula, the two-component core of the radio galaxy Centaurus A and the Moon at 13 cm with an East–West resolution of $2'$. Together with observations at other wavelengths, these data testify to the non-monotonic wavelength dependence of the integrated degree of linear polarisation of the northeastern component, and the large difference in the rotation measures of the core components of Centaurus A.

2.3.4 Studies of the Polarisation and Spectrum of the Galactic Radio Emission

After Razin's detection of the linear polarisation of the Galactic radio emission in 1955–1956 (at the Zimenki Laboratory, at wavelengths of 1.45 and 3.3 m), the main polarisation studies were conducted at the Staraya Pustyn' site (Razin, Khrulev, A. A. Mel'nikov, Popova, Vinyaikin, Teplykh and others).

Observations of the linear polarisation of the Galactic radio emission are an effective means of studying the physical conditions in the Galaxy. The angular distribution of the polarised radio emission on the sky and the frequency spectra of the polarisation parameters yield information about the distribution of ionised gas and relativistic electrons in interstellar space, as well as the structure of the Galactic magnetic field. With these goals in mind, studies of the angular distribution and frequency spectra of polarised regions in Loop III (Polar Region 147+8), the Northern Galactic spur and the vicinity of the North star were carried out, and the wavelength dependence of their angular sizes determined. A bright spot of linearly polarised radio emission with a polarisation temperature comprising 20% of the total temperature of the sky at 334 MHz was discovered in the direction of Polar Region 147+8 (declination $\delta = 60°$, right ascension $\alpha = 4^h 30^m$). In addition, the non-monotonic character of the frequency dependence of the degree of polarisation of the radio emission from this region was established.

Simultaneous measurements of the angular distributions of the linearly polarised and unpolarised Galactic radio emission at 334 MHz and interferometric measurements at 200 MHz demonstrated an anti-correlation between the temperatures of these two components, with the temperature of the total radio emission being proportional to the effective size of the Galactic disk. Thus, it was possible to determine directly and observationally that the linearly polarised radio emission is generated throughout the thickness of the Galactic disk.

The collected polarisation measurements obtained by NIRFI at 100–1000 MHz and data on the distribution of the brightness temperature over the sky at 85, 150 and 829 MHz show that the Galactic magnetic field out to distances of 3–10 kpc from the Sun in the sector with longitudes $l = 70$–$180°$ and latitudes $b > 0°$ is elongated along the Galactic plane and has a loop-like structure. The regular component of the field is directed toward $l = 55°$. The dispersion of the interstellar magnetic field in Galactic latitude is much less than unity, while the dispersion in longitude is equal to 0.5.

The anisotropy of the synchrotron radiation of the interstellar medium in a magnetic field with this structure can explain the distribution of the brightness temperature on the sky using a simple model with a radiating disk in the form of an ellipsoid of rotation with semi-axes of 14 and 0.8 kpc (without an intense radio halo).

The contribution of the radio halo at 85 MHz in the direction of the Galactic North pole is no more than 15% of the total sky temperature, and decreases with increasing frequency. Analyses have shown that no bright, well defined radio arms are observed in the sector $l = 70$–$180°$.

A series of studies of the spectrum of the non-thermal radio emission at 6.3–375 MHz was carried out in 1960–1980 (Getmantsev, Razin, Tarasov, Yu. V. Tokarev and others). These data refer to large-scale radio structures of the Galaxy, and can provide information about sources, accumulation volumes and the motion of cosmic rays, as well as about the intensity and spectrum of the metaGalactic radio emission. The efficiencies of various mechanisms for producing non-thermal cosmic radio emission have also been analysed.

2.3.5 Studies of the Cosmic Microwave Background

Proving the cosmological origin of the cosmic microwave background radiation required measurements of its spectrum in order to establish how well it corresponded to the spectrum of a perfect blackbody with a temperature of approximately 3 K. Investigations carried out at NIRFI were of prime importance for this problem (Kislyakov, Stankevich and others). A method based on using the radio emission of cooled absorption standards was developed as a means of obtaining accurate absolute measurements of the background radiation. This method was used to measure the temperature of the cosmic microwave background at various wavelengths: 2.55 mm (1968), 8.2 mm (1967, in collaboration with A. E. Salomonovich of FIAN), 3.2 cm (1966), 8.9 cm (1969) and 15, 20.9 and 30 cm (1968). At some wavelengths, the NIRFI data were the first obtained. The method of blocking the cosmic background radiation with the Moon was applied to obtain measurements at longer wavelengths, and the corresponding observations carried out by Stankevich at 47 and 73 cm using the 64-m Parkes radio telescope (Australia). These measurements showed that the Rayleigh–Jeans part of the spectrum was in full agreement with the spectrum of a blackbody with a temperature of 2.7 K. A high degree of small-scale isotropy (to 0.01%) was also demonstrated. These measurements were obtained at 11 cm with a resolution of 8′ on the 64-m Parkes radio telescope.

2.4 Studies at Millimetre Wavelengths

Radio astronomy studies at short millimetre wavelengths ($\lambda < 4$ mm) were begun at Gorkii by Kislyakov in 1954. By 1960, a modulation radiometer with a superheterodyne receiver operating at 4.1 mm had been devised at NIRFI, as well as a broadband (3–7 mm) detector radiometer. Kislyakov used these radiometers to obtain data on the phase dependence of the lunar radio emission at 4.1 mm and the spectrum of the solar radio emission at 3–7 mm. A method for calibrating millimetre-wavelength radio telescopes using the radio emission of the atmosphere taking into account the specifics of this wavelength range was developed, and detailed studies of the dependence of the atmospheric absorption at 4 mm on the meteorological parameters and the elevation of the observing site above sea level conducted (Kislyakov, Nikonov, Strezhneva).

Further, a unique series of detector radiometers operating at 1.8 mm (Naumov), 1.3 mm, 0.87 mm and 0.74 mm (Yu. A. Dryagin, Fedoseev) were developed at NIRFI, which were used for observations of the Moon and Sun, as well as studies of the atmospheric absorption of radio waves (L. M. Kukin, L. V. Lubyako, Fedoseev). At this stage, data on the phase dependence of the lunar radio emission over the entire interval of wavelengths from 0.8 to 4 mm were obtained. These measurements showed the adequacy of a one-layer model for the lunar surface to describe variations of its radio emission during lunation.

Studies of the radio emission of the Sun and Moon at 4 mm were then continued in collaboration with FIAN, using the FIAN 22-m radio telescope and instruments provided by NIRFI. These results showed the radiometric homogeneity of the equatorial band of the Moon, while at the same time differences in the physical properties of the upper layers of the lunar seas and continents were detected (Kislyakov, B. Ya. Losovskii, A. E. Salomonovich). The first high-resolution observations of sources of the S component on the Sun simultaneously at 4 and 8 mm were carried out (Kislyakov, Salomonovich). In this same period, Kislyakov, A. D. Kuz'min and Salomonovich conducted the first investigations of the radio emission of Venus at 4.1 mm (see Essay 1).

By 1965, the first broadband radiometers based on superheterodyne receivers with microwave intermediate frequencies in the Soviet Union had been devised at NIRFI (Kislyakov, Yu. V. Lebskii, Naumov). This made it possible to substantially increase the sensitivities of radiometers, and to turn to a new series of studies of the Moon, planets and discrete radio sources. Further improvement of millimetre and submillimetre radiometers was based on original ideas for the application of quasi-optical tracts (Fedoseev, Yu. Yu. Kulikov).

Studies of the solar radio emission at 1.35–1.7 mm on the 22-m radio telescope of the Crimean Astrophysical Observatory yielded data on the spectrum of the S component and protuberances at these wavelengths, as well as some information about the manifestations of solar flares (Kislyakov, Moiseev, V. A. Efanov, Fedoseev, Naumov, Lebskii and others). Spectra of the radio emission of Venus, Mars, Mercury and Jupiter obtained in collaboration with researchers at the Crimean Astrophysical Observatory were used to draw some conclusions about the conditions in the atmospheres and on the surfaces of these planets (Kislyakov, Efanov, Moiseev, Naumov, V. N. Voronov, I. I. Zinchenko and others). In another collaboration with the Crimean Astrophysical Observatory, a survey of discrete radio sources at 4 mm was carried out, which was the first in the Soviet Union and the most complete survey at that time. The spectra of a number of galactic sources, quasars and radio galaxies were continued to short millimetre wavelengths.

The construction of the RT-25×2 radio telescope made possible a survey of dark nebulae at several millimetre wavelengths, with the aim of detecting continuum emission. These searches yielded positive results in some cases (Kislyakov, Chernyshev, Zinchenko, A. A. Shvetsov and others). The existence of condensations in dark Galactic nebulae was later confirmed by millimetre observations with the Crimean 22-m telescope, and was also indirectly supported by the results of observations in lines of isotopes of the CO molecule (Kislyakov, B. E. Turner, M. A. Gordon). These last observations were carried out in collaboration with researchers at the National Radio Astronomy Observatory (USA) using one of the NRAO telescopes. New lines of this molecule were discovered and the presence of cyanamide at the Galactic centre demonstrated (Kislyakov, Turner, H. S. List, N. Kaif). The continuum emission from condensations in dark nebulae corresponds to emission from an optically thin absorbing layer (apparently dust).

The RT-25×2 radio telescope was used to study the brightness distribution over the disk of the quiet Sun, including limb effects, at wavelengths of 4.1, 6 and 8 mm, as well as chromospheric granulation, using simultaneous observations at 1.35 and 4.1 mm (Kislyakov, Chernyshev, Fedoseev, S. A. Pelyushenko and others).

Beginning in 1977, radio astronomy studies at short millimetre and submillimetre wavelengths were continued at the Institute of Applied Physics of the USSR Academy of Sciences, which was organised in Gorkii on the basis of several subdivisions of NIRFI. This period is characterised by a transition from spectroscopy with only crude frequency resolution to spectroscopy with high (10^{-5}–10^{-6}) relative frequency resolution. Multi-channel spectrometers operating at wavelengths near 2 and 3 mm with stabilised heterodynes based on backward-wave tubes were devised at the Institute of Applied Physics (A. B. Burov, Voronov, A. A. Krasil'nikov, Lebskii, N. V. Serov). This enabled the first observations in the USSR of interstellar molecules such as HCN (Burov, Zinchenko and others) and CO, in collaboration with the Bauman Technical University (Burov, Kislyakov, A. A. Parshchikov, B. A. Rozanov and others). The detection of HCN lines from dark nebulae led to a revision of lower limits for the density of hydrogen in these objects.

The application of multi-channel spectrometers in atmospheric research made possible studies of telluric lines in the rotational spectrum of ozone at wavelengths of 1.3–3.3 mm, which began at NIRFI (Kulikov, Fedoseev, Shvetsov, V. G. Ryskin and others). The possibility of remote monitoring of the total content and vertical profile of ozone was demonstrated, including during night-time and in the presence of light clouds. Radio astronomy methods for measuring the atmospheric absorption of radio waves were also used in systematic observations, whose results form the basis for making prognoses of the magnitudes of signals in Earth–Space communication sessions (Kislyakov, Fedoseev and others). This last work was carried out in collaboration with the Institute of Radio Physics and Electronics (V. F. Zabolotnii, Zinchenko, I. A. Iskhakov, A. V. Sokolov, E. V. Sukhonin, Chernyshev).

2.4.1 Development, Study and Application of Radio Astronomy Methods

Radio Interferometry with Ultra-high Resolution The first work on the creation of instrument complexes for radio interferometers with independent receiving systems designed to yield high resolution was begun at NIRFI in 1965 at the initiative of Troitskii. During 1965–1981, instruments operating at frequencies of 6, 9, 25, 86, 327, 408, 5300 and 22235 MHz were devised (Troitskii, Nikonov, Krotikov, V. A. Alekseev, E. N. Gatelyuk, A. E. Kryukov, B. N. Lipatov, A. S. Sizov, A. I. Chikin, M. V. Yankavtsev and others). Astrophysical investigations using large-scale Soviet radio telescopes began in 1969. The first measurements of the angular size of Cassiopeia A at decametre wavelengths were made; an angular resolution of 10^{-3} arcminute was realised in measurements of the angular sizes of cosmic masers at 1.35 cm with a baseline of 1100 km; interferometry of cosmic masers was used to synchronise time standards in different locations. This work was carried out in collaboration with the Lebedev Physical Institute, Space Research Institute, Institute of Radio Physics and Electronics, Crimean Astrophysical Observatory and Byurakan Astronomical Observatory.

The development of a methodical and technical basis for a new scientific research direction—precision radio astrometry using Very Long Baseline Interferometry (VLBI)—led to the proposal of the following methods: (1) aperture-frequency synthesis for obtaining radio images of cosmic radio sources with very high angular resolution; (2) differential long-baseline radio interferometry for establishing celestial and terrestrial coordinate systems, measuring the rate of rotation of the Earth and the motion of the poles, and studying tides in the Earth's crust, precessional–notational motions and tectonic and seismic phenomena. The principles of space radio astrometry were also considered. The potential accuracy of these methods when applied with VLBI is more than two orders of magnitude higher than the accuracies that can be achieved by other measurement methods.

Radio Astronomy Methods for Studying Antennas Radio astronomy methods for determining the energetic parameters of antennas were first proposed as early as 1955 by researchers at NIRFI and FIAN. Further, as a result of investigations conducted at NIRFI, these methods were developed and applied to complex antenna systems, including interferometric and phasometric systems (Troitskii, Tseitlin, Dmitrenko, Dugin and others).

In the 1960s, radio astronomy and radiometer methods using both cosmic radio sources and the radio emission of absorbing and scattering surfaces were developed at NIRFI. The most accurate antenna characterisations were achieved using the radio emission of "black" disks placed in both the Fraunhofer and Fresnel zones of the antenna—a method developed at NIRFI (Troitskii, Tseitlin).

Measurements of the radio fluxes of primary and secondary standard sources obtained at NIRFI enable the determination of the parameters of a whole range of antenna types with unprecedented accuracy (to within 5%). The application of these methods in industrial facilities during the development and production of antennas, and also in other types of facilities during the exploitation of antenna systems, proved to be very effective. Radio astronomy and radiometer methods for antenna measurements have a number of advantages in many applications, and sometimes provide the only means possible to obtain such measurements.

On the basis of work done at NIRFI, methods for characterising antennas are being successfully developed in a number of organisations in the USSR.

Radio Astronomy Methods for Analysing the Parameters of the Troposphere
In 1948–1952, methods for separately measuring tropospheric absorption by water vapour and oxygen based on the radio emission of the troposphere were proposed and developed at NIRFI (Troitskii, Tseitlin). Over the last 25 years, the most complete studies in the world of the absorption and emission of radio waves in the troposphere from millimetre to metre wavelengths have been carried out (Troitskii, Zhevakin, Tseitlin, Razin, Plechkov, Kislyakov, Stankevich, Khrulev, Dmitrenko, A. V. Troitskii, Naumov and others). Absorption and emission radio spectra of the troposphere at these wavelengths have been obtained, and their dependences on meteorological conditions analysed, which is very important for radio communications,

radar and other applications. The work in this area carried out at NIRFI also stimulated radiometer studies of the troposphere in other organisations in the Soviet Union.

As a result of many years of such tropospheric investigations, the possibility of determining the temperature, total water content and humidity of the troposphere by analysing tropospheric radio emission received simultaneously at several frequencies was demonstrated. Over the last 15 years, radiometer methods for obtaining tropospheric measurements, algorithms for solving inverse problems and principles for the construction of receivers have been developed. This led to the creation of a multi-channel radiometer system enabling the operative and continuous reception of information about the physical state of the atmosphere: profiles of the temperature and humidity, the total contents of liquid water and water vapour and the altitudes of clouds.

The developed instruments and methods for remote atmospheric measurements made it possible to conduct important meteorological studies in tropic zones of the Atlantic and Pacific Oceans and in regions within the polar circles, and to investigate the spatial (mesoscale) and temporal (hourly) variations in the total mass of water vapour and the water content of clouds.

Radio Astronomy Polarisation–Faraday Methods for Ionospheric Investigations (Razin, Teplykh, Popova) One of the important characteristics of the ionosphere is the total electron content along the line of sight (N_{II}). There exist several methods for measuring N_{II}, among which the most accurate and efficient are measurements of the Faraday rotation of the plane of polarisation of linearly polarised radio emission from sources located beyond the ionosphere. As a rule, geostationary satellites are used as sources for this purpose. Radio astronomy polarisation–Faraday measurements, which make use of the radio emission of polarised regions of the Galaxy, are very promising in connection with their economy and simplicity.

The first measurements of N_{II} using the radio astronomy polarisation–Faraday method were carried out in 1970 at the Staraya Pustyn' radio astronomy station. These measurements were carried out regularly at a wavelength of 1 m beginning in 1976. The sources used were a polarised region with declination $\delta = 61°$ and right ascension $\alpha = 4^h 30^m$ and the region of the North star. Numerous data on the behaviour of N_{II} during the night near the solar-activity maximum have been accumulated up to the present; the episodic appearance of a relative night-time maximum of N_{II} has been demonstrated and the statistics of this behaviour analysed.

When the manuscript for this collection was already at the publishers, we learned of the sudden death of the eminent Soviet scientist and teacher, professor and doctor of technical sciences Mikhail Mikhailovich Kobrin (Fig. 2.10)—one of the oldest radio astronomers in the Soviet Union and also one of the editors of this collection. M. M. Kobrin was one of the organisers of NIRFI, and was the Assistant Director of Scientific Research at NIRFI for many years, then the Assistant Head of his Department. He was the Dean of the Radio Physical Faculty of Gorkii State University, and the Science Rector of this university. The fruitful scientific activity of M. N. Kobrin in the area of radio astronomy is reflected in this essay. His contribution to

Fig. 2.10 Mikhail Mikhailovich Kobrin (1918–1983)

radio astronomy studies of the Sun was especially large. He was the Chairman of the Solar Radio Emission Section of the Scientific Council on Radio Astronomy of the USSR Academy of Sciences for many years, and did much to unite the efforts of Soviet radio astronomers in the development of this important research direction. He also personally took part in the development of radio astronomy in the German Democratic Republic and in Cuba.

Chapter 3
The Development of Radio Astronomy at the Sternberg Astronomical Institute of Lomonosov Moscow State University and the Space Research Institute of the USSR Academy of Sciences

L.M. Gindilis

Abstract This chapter provides information about the emergence and development of radio astronomy at the Sternberg Astronmical Institute of Moscow State University (GAISH), and further at the Space Research Institute (IKI). The main results of theoretical studies of mechanisms for the Sun, Galactic and extragalactic radio emission and their relationship to physical processes in space are laid out in detail. The results of observations carried out at the initiative of and with the participation of radio astronomers from GAISH and IKI using many radio telescope in the Soviet Union and abroad are also considered, including methods for space radio astronomy.

3.1 The Beginning of Radio Astronomy Studies at GAISH. Creation of the Department of Radio Astronomy

The establishment and development of radio astronomy at the Sternberg Astronomical Institute of Lomonosov Moscow State University (GAISH) and the Space Research Institute of the Academy of Sciences (IKI) is inseparably linked to the name of I. S. Shklovskii. The first radio astronomy research of Shklovskii was carried out at the end of the 1940s, when the results of radio observations of the Sun and Galaxy obtained in a number of countries during the period of the Second World War became known. Before this, his scientific interests were associated with classical problems in astrophysics, primarily spectroscopy.

In 1946, Shklovskii investigated mechanisms for the generation of the radio emission of the quiet Sun. This work, together with the work of V. L. Ginzburg and D. Martin, enabled the construction of an isothermal model for the solar atmosphere, which still lies at the basis of the theory of the radio emission of the quiet Sun. This model was used to draw the fundamental conclusion that the sources of solar radio emission are in the outer layers of the solar atmosphere—the chromosphere and corona—and not in the photosphere, with the main contribution to the solar emission at metre wavelengths being made by the corona. This conclusion was brilliantly confirmed by observations during the solar eclipse of May 20, 1947 (see

Essay 1). Simultaneously, Shklovskii put forth the hypothesis that the radio emission of the "perturbed" Sun was associated with plasma oscillations in the corona, which arose during the passage of flows of particles. This brave hypothesis, based on the closeness of the Langmuir frequency for plasma oscillations in the corona to the frequency of metre-wavelength radio waves, was further developed by V. V. Zheleznyakov (at the Radio Physical Institute in Gorkii), and is one of the main mechanisms lying at the basis of the theory of the radio emission of the perturbed Sun.

Shklovskii's monograph *The Solar Corona* (Moscow, Leningrad: Gostekhizdat, 1951) came out in 1951. The theory of the radio emission of the solar corona was developed by Shklovskii in close connection with problems associated with optical spectroscopy of the corona. This led to the hypothesis that the solar corona was ionised, and the basis of modern concepts of the hot corona. The idea that the corona is hot now seems completely obvious, but, in the 1950s, this concept had to be defended in a very stubborn struggle with adherents of the theory that the corona was cool. Based on his ionisation theory, Shkovskii determined the chemical composition of the solar corona and predicted the existence of hard electromagnetic radiation from the Sun.

Among works related to solar radio astronomy, we should mention the important research of Shklovskii and S. B. Pikel'ner at the beginning of the 1950s on the radio brightness distribution over the disk of the Sun. They showed that an enhancement in the brightness of the disk toward the limb should be expected at decimetre wavelengths, in contrast to the well known limb darkening observed in the optical. This predicted effect was confirmed observationally. The refraction of low-frequency radio waves in the corona was studied In the process of these investigations.

Shklovskii also made a fundamental contribution to the development of Galactic radio astronomy. In 1948, after van de Hulst had published his paper suggesting the fundamental possibility of observing the 21-cm radio line of hydrogen, Shklovskii carried out the necessary calculations of the transition probability for this line and the expected intensity of Galactic radio emission at the corresponding frequency. He showed that the detection of the line should be possible with the observational equipment available at that time. In 1951, the 21-cm radio line of hydrogen was detected nearly simultaneously in the USA, England and Australia.

Over the next several years (1948–1953), Shklovskii conducted a number of studies in which the foundations of radio spectroscopy of the Galaxy were laid. First and foremost, he considered the radio line of deuterium at 91.6 cm, as well as two radio lines of nitrogen at 15 and 30 m wavelength with negligibly small transition probabilities. Attempts to detect these lines have thus far not yielded clear results. Shklovskii then arrived at the fundamentally important conclusion that it should be possible to observe radio lines of interstellar molecules. He calculated the frequency and intensity of the 18-cm hydroxyl line. At that time, the available observational equipment was not able to detect this line, and it was finally detected only in 1963.

This discovery had huge importance, since cosmic maser emission at the frequency of the hydroxyl (OH) line was soon detected, and the sources of this maser emission turned out to be closely associated with star-formation processes (which was first indicated by Shklovskii in 1966). Together with his hydroxyl calculations, Shklovskii performed calculations for the CH line at 9.45 cm, which proved to be very weak, and was detected only in 1973. Soon after this discovery came a period of vigorous blossoming of molecular radio spectroscopy, which became one of the most important areas in radio astronomy.

At the beginning of the 1950s, Shklovskii carried out a series of theoretical studies on the thermal radio emission of the Galaxy. He proposed to observe such thermal radio sources (H II regions) on the radio telescopes of the Crimean Station of the Lebedev Physical Institute (FIAN), but the sensitivity of the available equipment was insufficient to realise this idea. Further, such observations were carried out successfully at the Kaluga Station of FIAN (see Essay 1).

Shklovskii's studies of the nature of the non-thermal radio emission of the Galaxy proved to be more fruitful. In 1952, he predicted the presence of a spherical component to the Galactic radio emission, which formed a so-called Galactic corona. A detailed theory of the radio corona of the Galaxy was developed by Pikel'ner and Shklovskii in 1957. The detection of the radio corona played a large role in the understanding of the mechanisms generating the non-thermal radio emission of the Galaxy, since it disproved the hypothesis that this emission was associated with "radio stars." Subsequent, more modern observations showed that the Galactic corona does not make as large a contribution to the overall radio emission of the Galaxy as was first proposed by Shklovskii.

Shklovskii actively took part in laying the basis for and applying the theory of radio synchrotron radiation. The idea that the non-thermal component of the cosmic radio emission was radiation by relativistic electrons in weak magnetic fields was expressed in 1950 by K. Kipenheuer, H. Alfven and N. Herlofson. In the early 1950s, a quantitative theory of synchrotron radiation was developed in the works of Ginzburg and his students and collaborators (I. S. Syrovatskii, A. A. Korchak, G. G. Getmantsev, V. A. Razin). In Shklovskii's view, a serious difficulty in applying this theory to the non-thermal radio emission of the Galaxy was the fact that (as it seemed at that time) the magnetic fields should be associated with clouds of interstellar gas, which lie in the Galactic plane, while the non-thermal radio emission seemed to be coming from a spherical system. However, after Pikel'ner showed in 1952 that the magnetic field is not localised only in interstellar clouds, but also in the intercloud medium, this difficulty was removed. Already in 1953, Shklovskii turned from the radio-star theory, which he had previously supported, and actively took part in the development of the theory of synchrotron radiation, concentrating his efforts on possible astronomical manifestations of the synchrotron mechanism. First and foremost, he invoked this mechanism to explain the radio emission of the Galactic corona, and showed that this corona can be treated like a reflection of the distribution of the electron component of cosmic rays. However, the question of the origin of the cosmic rays then arose.

In 1953, Shklovskii applied the synchrotron theory to estimate the energetics of supernova remnants. He showed that each supernova explosion gave rise to relativistic particles with a total energy of about 10^{41} Joules (about 10^{48} ergs). To estimate the efficiency of this process in generating cosmic rays, it was necessary to know the supernova rate in the Galaxy. Shklovskii's interest in historical chronicals making reference to supernovae dates to this time. He was the first radio astronomer to turn attention to a number of supernovae (such as the supernova of 185 in Centaurus, the supernova of 1006 in Lupus and others), and proposed to search for radio sources at the locations of these supernovae, which were indeed detected. By using data on supernovae referred to in historical chronicles, Shklovskii was able to increase previous estimates of the supernova rate by nearly an order of magnitude. It turned out that supernova explosions can fully compensate the decrease in the energy of cosmic rays due to nuclear collisions.

Very important studies of Shklovskii concerning the nature of the emission of the Crab Nebula occurred in this same period. At the beginning of 1953, he explained the radio emission of the Crab Nebula as being due to the synchrotron mechanism. However, the nature of the optical radiation remained unclear. The proposed theory of R. Minkowski that this was thermal radiation associated with free–free transitions encountered great difficulties. Shklovskii approached this problem from a completely unexpected direction: if the radio emission of the Crab Nebula was not a continuation of its thermal optical emission, might the optical continuum of the nebula represent a continuation of the radio synchrotron radiation to the optical? Subsequent calculations fully supported this insightful guess. Thus, Shklovskii was the first to develop the concept that there was a single radiation mechanism acting from the optical to the radio. The prediction of I. M. Gordon based on this theory that the Crab Nebula radiation should be appreciably polarised was soon confirmed observationally by the Soviet astronomers M. A. Bashakidze and V. A. Dombrovskii. Shklovskii's studies of the nature of the radiation of the Crab Nebula gave a push to the wide application of the synchrotron mechanism to other astrophysical objects.

Shklovskii's book *Radio Astronomy* came out in 1953 (Moscow: Gostekhizdat, 1953). Although formally this was considered to be a popular-science book, the main methods and achievements of radio astronomy at that time were laid out at a high scientific level. The book became an indispensible aid for many beginning radio astronomers.

In 1953–1954, Shklovskii gave the first course in radio astronomy in the Soviet Union in the Astronomy Department of Moscow State University, which was attended not only by students, but also by researchers at a number of institutes in Moscow who were starting to work in radio astronomy. A student seminar on radio astronomy under the direction of Shklovskii and A. E. Salomonovich (Shklovskii led the theoretical part of the seminar and Salomonovich the experimental part) was also organised in the Astronomy Department. This seminar, in which radio physicists and astronomers took part, was a very good school; by representing different approaches to radio astronomy (by astronomers and radio physicists), it facilitated understanding between the various types of scientists involved. The first participants in the seminar included such now well known radio astronomers as Yu. N. Pariiskii, N. S. Kardashev and N. S. Soboleva. Thus, fertile ground for the serious development of radio astronomy studies was created at the Sternberg Astronomical Institute.

The work of Shklovskii led to the development at GAISH of a new research direction—radio astronomy. A group of young astrophysicists specialising in radio astronomy formed around the scientist. Precisely this group formed the basis for the Department of Radio Astronomy formed at GAISH in 1953 under the supervision of Shklovskii. The first employees of the department were B. M. Chikhachev, who was at GAISH part time (his main work was at FIAN) and the fourth-year student in the Faculty of Mechanics and Mathematics of Moscow State University N. S. Kardashev. In 1953, V. N. Panovkin was accepted as a PhD student in radio astronomy under Shklovskii (his second supervisor was V. V. Vitkevich), with P. V. Shcheglov following a year later, in 1954. Also in 1954, Kardashev took part in the work of the Kaluga Station of FIAN under the supervision of N. L. Kaidanovskii, carrying out studies on the detection of discrete radio sources at 3 cm. These were the first observations of thermal nebulae at very short (for that time) radio wavelengths, and led to the detection of thermal radio emission associated with free–free transitions from diffuse gaseous nebulae (see Essay 1).

The development of radio astronomy investigations at GAISH required the establishment of an experimental base. With this goal in mind, an engineering group was formed in the Department of Radio Astronomy, and work was begun on the development of equipment, first and foremost a radiometer designed to operate at a wavelength of 21 cm (for both line and continuum observations). The engineers Yu. V. Bobrov, V. V. Golubev, V. I. Protserov and the laboratory assistants Kardashev, V. N. Panov and K. I. Petrova participated in the creation of this receiver. The work was carried out under the supervision of Chikhachev, with astrophysical supervision from Shklovskii. In those years, there was very close collaboration between the radio astronomy departments of GAISH and FIAN, which no doubt facilitated success in the establishment of radio astronomy research in both institutions.

3.2 The First Ten Years of the Department of Radio Astronomy (1954–1964)

The beginning of the first decade of work in the Department of Radio Astronomy was marked by big scientific events. On March 9–12, 1955, the fifth meeting on questions of cosmogony, dedicated entirely to problems in radio astronomy, took place in the conference hall of the Sternberg Astronomical Institute. In essence, this was the first nationwide meeting on radio astronomy.

In the following years, the Department of Radio Astronomy filled up with new members. V. G. Kurt joined the department in 1955, and V. I. Moroz in 1956, followed somewhat later by G. B. Sholomitskii, T. A. Lozinskaya, G. S. Khromov, V. N. Kuril'chik, M. I. Pashchenko and others. This was the first generation of researchers in the department, most of whom were students of Shklovskii.

Solomon Borisovich Pikel'ner (Fig. 3.1) was invited to the Astrophysics Faculty of Moscow State University as a professor in 1959. He had graduated from the Astronomical Section of Moscow State University in 1942, and his first teacher

Fig. 3.1 Solomon Borisovich Pikel'ner (1921–1975)

in his student years was none other than Shklovskii. After finishing his PhD and defending his thesis, Pikel'ner worked for more than ten years at the Crimean Astrophysical Observatory, in close collaboration with the important Soviet astronomy G. A. Shain. Having deeply imbibed of the tradition of his school, Pikel'ner himself became an eminent astrophysicist of our time. A person of unusual talent and rare personal qualities, he exerted a huge influence on a whole generation of Soviet astronomers. The many-sided scientific activity of Pikel'ner included close contact with GAISH Department of Radio Astronomy.

Pikel'ner was a theoretician with a very broad profile. He was one of the first to understand the importance of magnetic hydrodynamics for astrophysics. The main directions of his scientific work—cosmic electrodynamics, cosmic gas dynamics, plasma astrophysics—inevitably overlapped with problems in radio astronomy. Pikel'ner is associated with a number of important studies directly in the field of radio astronomy.

A large place in Pikel'ner's studies was occupied by the physics of gaseous nebulae, including supernova remnants. In 1956, he pointed out that the life times of the relativistic electrons in the Crab Nebula are appreciably shorter than the life time of the nebula itself, and proposed the presence of continuous electron acceleration in the Crab. The joint work by Pikel'ner, Ginzburg and Shklovskii in which they proposed a method for estimating the magnetic field in the Crab Nebula based on the location of the break in the nebula's radio spectrum dates to this same period. This method was further widely applied, and continues to be applied, in radio astronomy to estimate the ages and magnetic fields of various objects.

At the beginning of the 1960s, the idea that the interstellar gas was heated and ionised by ultraviolet radiation from stars was predominant. However, the observations of Faraday rotation of the plane of polarisation of cosmic radio sources that were known by that time did not agree with this picture. There were also difficulties in explaining the temperatures of neutral-hydrogen clouds implied by observations in the 21-cm line. Pikel'ner's analysis of these data showed that the interstellar medium was heated, not by ultraviolet radiation, but by soft cosmic rays.

Important theoretical studies were also carried out by Shklovskii in this period. In 1955, he showed that the optical emission of the jet emerging from the nucleus of the galaxy M87 is emission by relativistic electrons, and proposed a mechanism for the generation of these electrons. Although this particular work is concerned with optical emission, it was also of importance for radio astronomy, since it suggested that the idea that there was a single radiation mechanism acting in the radio and in the optical was applicable not only to Galactic, but also to extragalactic objects.

In 1953–1956, Shklovskii used synchrotron theory to estimate the energy (10^{51}–10^{53} J) of the relativistic particles (and magnetic fields) in radio galaxies. Somewhat later, similar calculations were carried out by G. Burbidge, who obtained the same results. This sharply posed the question of the source of the energy of radio galaxies.

Shklovskii's book "Cosmic Radio Emission" (Moscow, USSR Academy of Sciences) came out in 1956. This was the first monograph on radio astronomy in the Soviet Union, and one of the first in the world. A whole generation of Soviet radio astronomers were trained on it.

In 1960, Shklovskii predicted the secular variation in the radio flux from the source Cassiopeia A, based on the picture of this object as an expanding supernova remnant. Building on this idea, he constructed a diagram relating the radio brightness of a supernova remnant with its distance, which provided a new (radio astronomy) method for determining the distances to supernova remnants, which has been put to good use in studies of these objects. In that same year, Shklovskii developed a theory of old supernova remnants by applying a self-similar solution for a strong explosion in a medium with a constant heat-capacity. Based on this theory, he predicted the existence of soft X-ray emission from supernova remnants.

The discovery of quasars in 1963 stimulated a number of studies by Shklovskii dedicated to these important astrophysical objects. Already in 1963, he hypothesised the variability of the optical emission of quasars. Based on this suggestion, GAISH researchers Yu. N. Efremov and A. S. Sharov analysed old photographs of the quasar 3C273 and, indeed, found that its emission was variable. Further, in a series of studies in 1963–1964, Shklovskii showed that the chemical composition of quasars coincides with the normal (in other words, solar) composition. In 1965, he predicted that the radio emission of active galactic nuclei and quasars should also be variable, and developed theory associated with this, which was subsequently fully confirmed by observations.

N. S. Kardashev was very active beginning in the second half of the 1950s. His most important theoretical work in these years was a calculation demonstrating the possibility of observing radio recombination lines arising in transitions between highly excited states of the hydrogen atom. He reported on these results at the Xth Assembly of the International Astronomical Union, which took place in the Summer of 1958 in Moscow. A search for new radio lines began both in the USSR and at major radio telescopes of the USA. These lines were first detected by two groups of Soviet radio astronomers, in Pulkovo and FIAN (see Essay 1). There is no doubt that this was one of the greatest achievements of Soviet radio astronomy. It is especially gratifying to note that all stages of these studies (from posing the problem, to establishing its theoretical basis, to observationally detecting the lines) were carried out in the Soviet Union. The authors of this discovery (E. V. Borodzich, Z. V. Dravskikh, A. F. Dravskikh, N. S. Kardashev and R. L. Sorochenko) were awarded a diploma dated August 31, 1964 for this work.

Toward the end of the 1950s, the manufacture of a correlation radiometer operating at a wavelength of 21 cm was completed in the Department of Radio Astronomy. In 1960, a group of researchers under the supervision of Kardashev began observations with this radiometer on the RT-22 radio telescope in Pushchino. The 21-cm continuum radio fluxes of a large number of diffuse gaseous nebulae were measured. In addition, Kardashev estimated an upper limit for the 21-cm line radiation, in other words, for the content of neutral hydrogen, in clusters of galaxies in Corona Borealis and Gemini.

Together with the observations in Pushchino, the distribution of neutral hydrogen in our Galaxy was studied using the radio telescope of the Crimean Station of FIAN. This work was carried out in close collaboration with FIAN scientists (Sorochenko, Chikhachev). Kardashev, O. N. Generalov, Lozinskaya, Kuril'chik and Lekht took part in these observations, and the observational data were reduced by N. F. Sleptsova and Lozinskaya. The results were used to construct a relief map of the distribution of hydrogen in the Galaxy, characterising the thickness of its hydrogen layers. The magnitude of deviations of the gaseous disk from the Galactic plane for various distances from the Galactic centre, characterising the deforming (bending) of the gaseous disk relative to the Galactic plane, were also obtained (Kardashev, Lozinskaya). This effect had been detected earlier by Australian and Dutch researchers. The Crimean observations made it possible to appreciably refine the available data for the Northern hemisphere, and also to extend them to much larger distances from the Galactic centre.

In 1962, Kardashev studied in detail the nature of the spectra of sources of nonthermal radio emission taking into account various mechanisms for energy losses and gains. Based on theoretical calculations, he derived information about the character of such spectra (their shape, cutoffs and breaks) and the expected behaviour for their time variations. This work was very important for our understanding of the processes occurring in non-stationary sources. In that same year, a group of authors from GAISH and FIAN (Kardashev, Kuz'min, Syrovatskii) published an article generalising the results of observations of Cygnus A in the Astronomicheskii Zhurnal. They determined the break frequency in the radio spectrum of Cygnus A and derived the first estimate of the age of this radio source—about 0.5 million years.

In 1962, observations on the large antennas of the Deep-Space Communications Centre at 32 and 7 cm were begun. Sholomitskii investigated radio sources in clusters of galaxies. In 1964–1965, he discovered variability of the radio source CTA-102. Kuril'chik carried out detailed studies of the radio emission of normal galaxies, while Khromov was interested in the radio emission of planetary nebulae. The observations of the occultation of the quasar 3C273 by the Moon had extreme importance, since they made it possible to identify the presence of a very compact component. At various times, M. G. Larionov, Sleptsova and others took part in this work. Many GAISH radio astronomers acquired very good training in radio astron-

Fig. 3.2 G. B. Sholomitskii, I. S. Shklovskii and N. S. Kardashev (*left to right*) in the GAISH conference hall after the press conference about the radio emission of the source CTA-102 in 1965

omy observations in the Crimea, through the use of such modern and new equipment for that time as masers, parametric amplifiers and so forth.

Historically, the detection of variability of the radio emission of CTA-102 by Sholomitskii is of special interest. This source was selected because, based on its spectrum, Sholomitskii had suggested the possibility of secular variations of its radio flux, analogous to the variations of Cassiopeia A. At the same time, Kardashev, based on the hypothesis that this radio emission had an artificial origin, suggested the possibility that there might be periodic variations of the radio flux. The observations carried out by Sholomitskii in 1964–1965 confirmed the presence of periodic variations with a period of 102 days (Fig. 3.2). From an experimental point of view, this work was conducted with all due care: the flux of CTA 102 was measured (relative to the flux of the radio source 3C48), using as a control the source CTA 21 (which has a similar flux and is nearby on the celestial sphere), which did not show any radio-flux variations. All possible sources of error were investigated in detail and taken into account. Nevertheless, this result was met with a certain scepticism, in part due to the fact that it was completely unexpected and partly because the nature of the source had become associated with the hypothesis of extraterrestrial civilisations. The verification of the result by a number of observatories led to the discovery of a completely new and fundamental property of quasars—variability of their radio emission. However, the variability of the radio flux of CTA-102 was not confirmed. This effect was detected again only in 1972, by the Australian radio astronomer J. Hunstead, and then also by others. It is currently thought that the variability of this source has a transient nature, with intervals of variability and stability alternating in time.

In 1962, Sholomitskii considered the ratios of the radio fluxes of the lobes of double radio sources, as a possible consequence of Doppler effects acting during the expansion of these components. This idea was further developed by radio astronomers at Cambridge. In 1965, Kuril'chik pointed out the presence of strong evolutionary effects (variation of the radio luminosities of components with time), which he believed could play a dominant role in explaining the observed flux ratios.

Among other results obtained in the early 1960s, we should note the studies of synchrotron reabsorption by V. I. Slysh, which led to Slysh's well known formula for the estimation of the limiting angular sizes of sources of synchrotron radiation.

At the end of the 1950s, GAISH Radio Astronomy Department was actively involved in the programme of space investigations carried out in the Soviet Union. For example, Shklovskii established contact with S. P. Korolev and participated in the planning of many experiments. Kardashev and Slysh worked on the development of the first space radio telescope for observations of long-wavelength radiation (at frequencies below the critical frequency of the ionosphere).

The first successful measurements of long-wavelength radio emission at wavelengths of 150 and 1500 m using instruments created at the Sternberg Astronomical Institute were conducted on board the *Zond-2* and *Venera-2* spacecraft in 1964–1965.

3.3 The Search for Signals from Extraterrestrial Civilisations, and the RT-MGU and RATAN-600 Projects

To an appreciable extent, the development of radio astronomy at the Sternberg Astronomical Institute was held up by not having its own observational base. In spite of its large contribution to the development of radio astronomy, GAISH did not have its own radio telescope, and GAISH researchers were forced to carry out their observations using antennas that belonged to other institutions. This problem was partially addressed when GAISH participated in the equipping of the RATAN-600 radio telescope, which gave the right to continuous access to this instrument and the organisation of their own observational base there. The path that led to this solution was not trivial, and its beginnings were associated with a problem that falls outside the sphere of radio astronomy itself—the search for signals from extraterrestrial civilisations.

Interest in this problem grew at the end of the 1950s, at the dawn of space studies. Shklovskii and his colleagues began to discuss this question in the summer of 1961. The now widely known book of Shklovskii "The Universe, Life and Intelligence" (Moscow, USSR Academy of Sciences) came out in 1962, and subsequently had a large influence on the development of investigations of the question of extraterrestrial civilisations in the Soviet Union. Kardashev analysed the sending of information by extraterrestrial civilisations in detail. His main conclusion was that the level of technology used in ground-based radio astronomy should enable the detection of signals from highly advanced civilisations, provided only that they were located in our Galaxy, or even in neighbouring galaxies. This opened good perspectives for wide searches for such signals. In 1963, an initiative group was formed at GAISH (Kardashev, L. M. Gindilis, Slysh), which set the goal of drawing the attention of the scientific community to the problem of searching for signals from extraterrestrial civilisations and organising practical investigations with this aim. The overwhelming majority of scientists with whom this question was discussed considered the task to be quite reasonable. At the end of 1963, Shklovskii and Kardashev met with Ambartsumian, who suggested conducting a scientific meeting to provide a forum for multi-faceted discussions and estimating the state of this problem. This meeting took place in May 1964 at the Byurakan Astrophysical Observatory, and marked the beginning of studies of the search for extraterrestrial signals in the Soviet Union. The initiators of these studies were radio astronomers of Moscow State University.

The Byurakan meeting about studies connected with the detection of extraterrestrial signals included detailed analyses of radio sources in the centimetre and decimetre wavelength ranges that were most favourable for interstellar communications, with the aim of revealing possible artificial sources (according to the expected criteria for a signal being artificial). Kardashev put forward the idea of carrying out complete radio surveys of the sky at centimetre wavelengths. Naturally, this task was of considerable interest for purely radio astronomy purposes as well. Kardashev proposed the construction of a specialised meridian radio telescope operating at centimetre and millimetre wavelengths for this purpose. A Kraus-type radio telescope was adopted as the basis for this instrument. The development of a preliminary model of the radio telescope, which acquired the name "RT-MGU," was produced in 1964 in the Department of Radio Astronomy of GAISH. Kardashev, Slysh and Gindilis (who had transferred to the Radio Astronomy Department of the High-Mountain Station of GAISH) took part in this work. Substantial help in the development of the project was provided by P. D. Kalachev of FIAN.

The radio telescope was designed to operate at 0.4–10 cm. An antenna consisting of two reflecting surfaces—one parabolic and one planar—and a secondary mirror (feed) was envisioned. The horizontal reflecting suface was 414×8.2 m in size, and could rotate within 52° of the vertical, enabling coverage of a declination interval of 105° and making it possible to observe 80% of the entire sky at a latitude of 45°. The projected geometrical area of the antenna aperture was 2000 m^2, and the beam full-width at half-maximum size was $2.6'' \times 3.5'$ at 0.4 cm. The time for surveying the observable part of the sky was supposed to be 14.5 years at 0.4 cm and 0.5 years at 10 cm. The calculated minimum detectable fluxes at these wavelengths (for a bandwidth of 1000 MHz and a noise temperature of 100 K) were 0.25 and 0.05 Jy, respectively. The expected number of detected sources was of the order of several tens of thousands. In addition to carrying out the surveys, the telescope could be used to obtain high-accuracy measurements of the right ascensions and detailed one-dimensional maps of the source brightnesses. It was intended that the radio telescope could be outfitted by the end of the 1960s, and it would have had very good parameters for that time.

The cost of the planned RT-MGU radio telescope was two million rubles.[1] Unfortunately, the university was not able to allocate such a sum. The rector of

---

[1] This sum is sometimes mistakenly taken to be the first stated price of the RATAN-600 telescope.

Moscow State University, Academician I. G. Petrovskii, who had always been sympathetic with the needs of radio astronomers, discussed with Keldysh, the President of the USSR Academy of Sciences, the possibility of outfitting the radio telescope jointly with the Academy of Sciences on a share basis. This proposal was approved, and the matter was given over to the USSR Academy of Sciences Scientific Council on Radio Astronomy. Simultaneously, the Radio Astronomy Section of the Main Astronomical Observatory proposed to outfit a radio telescope based on a variable-shape antenna constructed earlier for the same wavelength range and having a similar effective area for its operational section. During the discussion of both projects, it was decided to combine the two. The main reflecting surface of the variable-shape antenna was supplemented with a flat reflector making it possible to survey the sky with a knife beam (see Essay 5). This is how the RATAN-600 radio telescope project came into being. Moscow State University's share in the project was determined to be 300,000 rubles. The funding provided by the university made it possible to initiate the financing of the project and carry out initial work on the telescope. (For more detail about the creation of the RATAN-600 radio telescope see Essays 5 and 6.) Shkovskii, Kardashev, Gindilis and Zabolotnii from GAISH took part in developing the project. Further, A. E. Andrievskii participated in the planning of the radio-receiver complex. In addition, the Assistant Director of the Sternberg Astronomical Institute, P. S. Soluyanov, was closely involved in all matters having to do with the outfitting of the RATAN-600 radio telescope. In order to facilitate the successful fulfilment of this work, the Division of General Physics and Astronomy of the USSR Academy of Sciences appointed Gindilis as the authorised representative of the Division in connection with the RATAN-600 project. Simultaneously, Andrievskii was established as an assistant head construction engineer in connection with the receiver apparatus.

Immediately after the decision was made to construct the radio telescope (1966), the RATAN-600 group under the supervision of Gindilis was formed at GAISH. It was supposed that work associated with the RATAN-600 would occupy a very important place in the activity of the GAISH Department of Radio Astronomy. However, the beginning of the establishment of the Department of Astrophysics of the Space Research Institute of the Academy of Sciences in 1967 (based on the GAISH Department of Radio Astronomy) substantially changed these plans. This left primarily the youngest members of the Department of Radio Astronomy in the RATAN-600 group: A. E. Andrievskii, A. G. Gorshkov, M. G. Larionov, V. K. Konnikova and V. V. Danilov.

The theme of scientific studies of the Sternberg Astronomical Institute using the RATAN-600 telescope was set from the very beginning—conducting complete surveys at centimetre wavelengths. The group was given the task of creating a receiver complex suitable for survey work. By the time the first stage of the construction of the RATAN-600 began (1974), the main stages of this work had been completed. In 1974, a GAISH laboratory was established at the RATAN-600 site, whose acting head was Gindilis, with Larionov replacing him in this capacity in 1976.

3.4 Establishment of the Department of Astrophysics at IKI. Radio Astronomy at GAISH in the Transitional Period (Second Half of the 1960s)

In 1966, Shkovskii was invited to organise and head the Department of Astrophysics in the newly established Space Research Institute (IKI) of the USSR Academy of Sciences. This opened broad possibilities, first and foremost in the development of space-related research, which was intensively carried out in the GAISH Department of Radio Astronomy at the end of the 1950s.

The administration of IKI supported the point of view of Shklovskii that space studies should be developed in close association with ground-based observations. During the creation of the radio astronomy departments at IKI, the need to retain close ties with the GAISH Department of Radio Astronomy was also recognised, which ensured good training of young specialists.

Radio astronomy studies at IKI were headed by Kardashev. A number of leading GAISH radio astronomers went to IKI together with him, such as Slysh (to organise work first in low-frequency radio astronomy, then radiospectroscopy) and Sholomitskii (to develop work in submillimetre astronomy), as well as some GAISH engineers. L. I. Matveenko was invited from FIAN, and headed the radio interferometry section. A number of graduates of the Astronomy Department of Moscow State University also came to IKI in those years (I. E. Valtts, V. A. Soglasnov, V. A. Soglasnova, V. M. Charugin), as well as M. V. Popov, who had obtained his PhD at GAISH. V. S. Etkin and I. A. Strukov, who had extensive experience in the development of high-sensitivity amplifiers and had earlier established close contact with the GAISH Department of Radio Astronomy, were invited in the same period, to develop radio astronomy equipment in the IKI Department of Astrophysics. As the formation of radio astronomy departments at IKI proceeded, scientific work was conducted in collaboration with the GAISH Department of Radio Astronomy.

By the end of the 1960s, the preparation of the first version of a radiometer operating at 3.5 cm was completed in the RATAN-600 group under the supervision of Andrievskii, jointly with the Laboratory of Problems in Radio Physics, whose researchers V. M. Mirovskii, E. E. Spangenberg and V. V. Nikitin took part (under the supervision of Etkin and Strukov). In 1968, this radiometer was installed on the RT-22 radio telescope of the Crimean Astrophysical Observatory, and a large series of measurements of the 8550-MHz fluxes of radio sources was carried out. A large group of researchers from GAISH, IKI, the Laboratory of Problems in Radio Physics and the Crimean Astrophysical Observatory participated in this work.

The preparation of an addition to the radiometer enabling transfer of the data to a computer was completed under the supervision of Larionov in 1969. In April–May 1969, this apparatus was used to carry out the first Soviet survey of a section of the sky at 3.5 cm on the Crimean RT-22 telescope. These observations were conducted with the aim of working out methods for future surveys on the RATAN-600, but, of course, they had other independent scientific value as well. Their analysis of these survey data (together with data from the Ohio State survey at 21 cm) led

Gorshkov and Popov to infer the existence of a new population of Galactic radio sources forming a spherical component of the Galaxy.

By the end of the 1960s, the manufacture of a spectral radiometer operating at 21 cm was also completed in the GAISH Department of Radio Astronomy, which was developed and constructed over many years under the supervision of Pashchenko. This put the question of carrying out observations with this instrument on the agenda. Shkovskii was able to get the director of the Observatoire de Paris–Meudon to agree to conduct joint studies using the large radio telescope at Nancay. This was a very fortunate arrangement, since the Nancay telescope was by far the most suitable for these investigations. Since observations of radio lines of OH were considered to be the most promising, the high-frequency part of the receiver was adjusted to enable observations at 18 cm. This collaboration with the Meudon Observatory was very fruitful. The equipment used was subsequently considerably modernised. The first observations at Nancay were carried out in 1969 (Slysh, Pashchenko, Lekht, Strukov). In the following years, from 1969 to 1978, researchers of GAISH and IKI conducted a series of very valuable studies of regions of OH emission using the Nancay telescope (see Essay 6). The spectral radio astronomy group of GAISH was formed during this work (Pashchenko, Lekht, Rudnitskii).

In that same period, work on equipment suitable for millimetre and submillimetre astronomy began at GAISH at the initiative of Kardashev. At first, this work was assigned to the Oimyakonskii Station of GAISH. Emission at submillimetre and far infrared wavelengths are absorbed by water vapour in the Earth's atmosphere. The most radical means to remove this limitation was to place an observing device outside the atmosphere. The absorption can be decreased by choosing a site on the Earth's surface where there is a low water-vapour content in the atmosphere. This condition is satisfied by high mountainous regions and regions with very low temperatures. A station was organised near the North Pole (the Oimyakon settlement, Irkutsk), to measure the atmospheric opacity in the infrared and submillimetre ranges and to evaluate possibilities for observations at these wavelengths. The apparatus used was constructed with the participation of the Moscow State Pedagogical Institute (E. M. Gershenzon, I. K. Morozov, Etkin). N. V. Vasil'chenko, Kardashev, Moroz, Morozov, Repin and Khromov took part in the work of this station. Observations were carried out in February 1966 at a temperature of −60°C. However, the water content of the atmosphere turned out not to be as low as had been expected (only an order of magnitude lower than at middle latitudes in the European part of the USSR in wintertime). This was due to the presence of a stable temperature inversion (in height). Analysis of these measurements indicated that regions in Eastern Siberia were promising for observations at 1 mm, but could not compete with balloon flights for measurements at wavelengths shorter than 300 μm.

Researchers at IKI continued to search for an optimal place for ground-based observations, leading to the selection of two locations: the mountains of southern Uzbekistan for millimetre observations, and in the high-mountain part of Pamira for submillimetre observations (see Sect. 3.5).

The development of techniques for measurements at millimetre and submillimetre wavelengths was worked on by Zabolotnii at GAISH and Sholomitskii at IKI.

A submillimetre radiometer based on an "electron bolometer" cooled by liquid helium to a temperature of 4.2 K was developed during 1967–1968. A series of observations at 1–2 mm wavelength were carried out with this radiometer installed on the Crimean Astrophysical Observatory RT-22 telescope in 1969–1975. Zabolotnii, Slysh, Soglasnova, Sholomitskii and others took part in these observations. In spite of the fact that the RT-22 surface was not designed for these wavelengths and the humidity was fairly high during the observations, a sensitivity of about 10 Jy was realised for an hour integration time. The brightness temperatures of all the giant planets of the solar system were measured at 1.4 mm, as well as the fluxes of a number of Galactic (the Crab Nebula, NGC 7027, W51, W49) and extragalactic (NGC 3034, 3C273 and others) sources.

In the second half of the 1960s, as in earlier years, intensive theoretical studies were conducted in the GAISH Department of Radio Astronomy. The year 1965 was marked by a number of important discoveries in radio astronomy. The most fundamental of these was the detection of the "relict radiation"—the cosmic microwave background radiation. The possibility of detecting this radiation was predicted by Novikov, who began this work at GAISH and is now working at the IKI Department of Astrophysics, and by A. G. Doroshkevich. In 1964, they calculated the entire spectrum of the radiation at all wavelengths from all sources in the Universe (radio galaxies, radio stars and so forth), taking into account their possible evolution, the expansion of the Universe and other cosmological effects. They showed that, if the Universe was indeed hot at the beginning of its expansion, the residual "relict radiation" at centimetre and millimetre wavelengths should be many orders of magnitude higher than the total radiation from discrete sources, and should be detectable. The cosmic microwave background radiation was discovered a year after by A. A. Penzias and R. W. Wilson (USA). The value of the work of Soviet scientists was noted in the Nobel lecture given by Penzias (see also Essay 4).

The term "relict radiation," which has become well established in the language of modern science, was proposed by Shklovskii. Later, in 1966, he proposed a method for determining the temperature of this radiation based on the intensity of optical molecular lines emitted by the interstellar gas (CN and other molecules).

In 1965, Kardashev calculated a model for the generation of the magnetic field of the neutron star (essentially a pulsar!) in the Crab Nebula. It is interesting that this work was carried out three years before the discovery of pulsars, and, of course, the very term "pulsar" did not exist at that time. Kardashev showed that the neutron star that formed after the supernova explosion of 1054 should have a very strong magnetic field. As the star rotates on its axis (with a period of less than one second), the magnetic field lines in the envelope are "wound up" by the rapid rotation of the neutron star. This field is the magnetic field of the Crab Nebula. This work of Kardashev has great importance for our understanding of the physics of the Crab Nebula and similar objects. The theory of the emission of the Crab pulsar was developed by Shklovskii in 1970.

3.5 Radio Astronomy Research at IKI

Radio astronomy research was methodically developed in the following directions: (1) space radio telescopes and methods for radio astronomy studies in space; (2) very long baselines radio interferometry; (3) studies of the propagation of radio waves in the interstellar medium; (4) analysis of fine temporal structure of rapidly variable processes using synchronous observing methods; (5) spectral studies; (6) submillimetre studies; (7) studies of the nature of radio sources; (8) searches for radio signals from extraterrestrial civilisations; (9) theoretical research.

Space Radio Telescopes and Methods for Radio Astronomy Studies in Space
Work on large radio telescopes has been conducted under the supervision of Kardashev beginning in 1967.

In 1968–1970, the possibility of realising an Earth–Moon interferometer was studied in the IKI Department of Astrophysics in collaboration with the Observatoire de Paris–Meudon. Kardashev, Slysh and Matveenko took part on the Soviet side, and Bloom, Denis, Leke and Steinberg on the French side. These studies (Matveenko et al.) showed that it was optimal to place the space radio telescope into an orbit around the Earth.

In 1970, Kardashev and Pariiskii demonstrated that space interferometers provide the unique possibility of obtaining three-dimensional images of radio sources and directly determining the distances to these sources. They also showed the possibility of obtaining direct measurements of the cosmological curvature of space and the main cosmological parameters, since the entire Universe would be located in the near field zone of an interferometer baseline with a length of several astronomical units (one AU is the distance from the Earth to the Sun). At the initiative of Ginzburg, the first broad discussion of the scientific and technical problems associated with space radio astronomy was conducted in December 1970.

The major specialists of the Institute of Project Construction A. G. Sokolov and A. S. Gvamichava, as well as V. I. Usyukin of the Bauman Moscow Higher Technical Institute, were recruited to develop methods for the construction of space radio telescopes. In subsequent years, new organisations and specialists in various scientific and technical fields who understood the importance and promise of constructing a radio telescope in space were attracted to this work.

In 1968–1978, many varied radio-telescope construction designs (with reflector sizes from 10 m to several kilometres) were analysed, together with methods for deploying the telescopes: inflatable elements, opening by centrifugal forces, the use of reactive motions, etc. In addition to options with some sort of automated deployment, constructions that could be deployed with the aid of cosmonauts were also considered. The direct supervision of all these studies was provided by Kardashev. A large amount of help in organising work on the construction of space radio telescopes was given by the President of the USSR Academy of Sciences Keldysh, and then A. P. Aleksandrov, as well as the Vice President of the USSR Academy of Sciences and Chairman of the Scientific Council on Radio Astronomy V. A. Kotel'nikov.

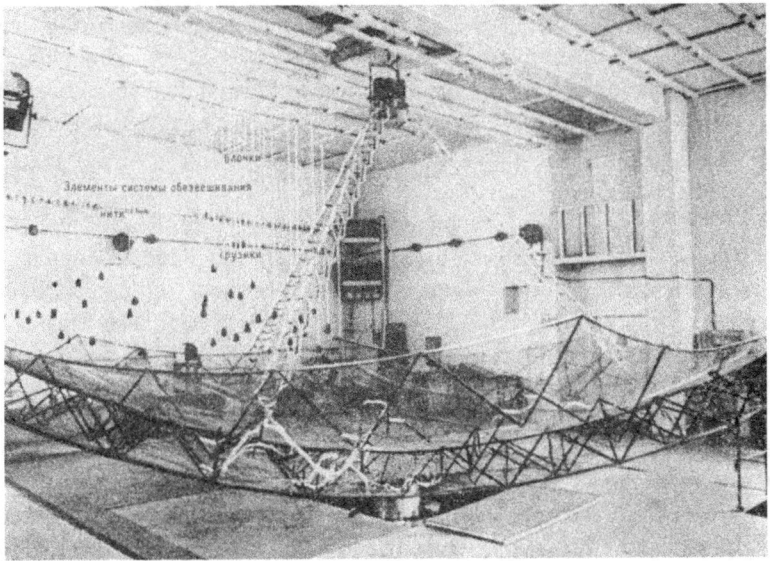

Fig. 3.3 10-m space radio telescope undergoing testing

These investigations met with success in 1979: the first space radio telescope—the KRT-10, with a diameter of 10 m—was constructed and put into orbit (Fig. 3.3). On June 30, 1979, the "Progress-7" cargo spaceship transported the KRT-10 components to the "Salyut-6" orbiting station. The cosmonauts V. A. Lyakhov and V. V. Ryumin assembled the antenna and focal container together with 12 and 72-cm receivers inside the adapter bay, and the low-frequency blocks and controlling device inside the instrument bay. On July 18, 1979, the "Progress-7" left the station, and the space radio telescope was moved out and deployed. During testing of the telescope, methods for radio astronomy observations in free space were developed for the first time. The Soviet scientists N. S. Kardashev, V. M. Arsent'ev, V. D. Blagov, A. S. Gvamichava, Yu. I. Danilov, v. I. Verzhatii, G. A. Dolgopolov, I. B Zakson, V. A. Krasnov, Yu. P. Kuleshov, I. M. Moshkunov and A. G Sokolov were awarded a State Prize for this work.

Substantial contributions to the construction of the space radio telescope were also made by V. I. Altunin, V. V. Andrianov, V. I. Buyakis, Sh. A. Vakhidov, L. A Gorshkov, V. V. Klimashin, V. I. Komarov, N. P. Mel'nikov, G. S. Narimanov, N. Ya. Nikolaev, M. V. Popov, V. A. Rudakov, A. I. Savin, R. Z. Sagdeev, Yu. P. Semenov, V. S. Troitskii, K. P. Feoktistov, G. S. Tsarevskii, I. S. Shklovskii, V. S. Etkin and other Soviet scientists.

Possibilities for realising long-term operation of space radio telescopes were investigated at IKI in the 1970s. More distant prospects for space radio astronomy were also studied (the possibility of constructing large modular antenna structures and putting them into very high orbits, with sizes of one astronomical unit or more). This could open possibilities for posing problems associated with astrophysics and the search for extraterrestrial civilisations on a completely new level. In particular, as was shown by Kardashev, Pariiskii and Yu. N. Umarbaeva (Special Astrophysical Observatory, USSR Academy of Sciences) in 1973, three space radio telescopes in high orbits could comprise a new type of instrument that could be used to determine the three-dimensional structure of a cosmic radio source by measuring the curvature of the incident radiation (the method of radio holography).

Very Long Baseline Radio Interferometry (VLBI) The idea of creating very long baseline radio interferometers using independent (unconnected) antennas was proposed in the early 1960s by Matveenko, Kardashev and Sholomitskii, and was published in the journal *Radiofizika* in 1965. This idea was not realised in a practical sense at GAISH, and this work was carried out at FIAN by Matveenko.[2] In 1969, Shklovskii invited Matveenko to IKI for the purpose of organising radio interferometry studies in the newly created Department of Astrophysics. In that same year, it was decided to move work on VLBI from FIAN to IKI. Matveenko headed the Radio Interferometry Section organised at IKI. Together with theoretical investigations of possibilities for creating space interferometers (see above), experimental studies using ground-based telescopes were begun in the Radio Interferometry Section. During the first years, while the necessary experimental base was being established at IKI, work on radio interferometry conducted using the equipment of FIAN and the Crimean Astrophysical Observatory.

The first Soviet–American experiment between Simeiz and Green Bank was realised in 1969, and a new experiment involving Simeiz, Green Bank and Goldstone in 1970. Observations were carried out at 2.8, 3.6 and 6 cm. As a result of these observations, the possibility of obtaining very high resolution (several tenths of a milliarcsecond) at centimetre wavelengths on intercontinental baselines was shown for the first time. Further, through the joint efforts of a number of countries, a global interferometric array was created, which includes the RT-22 telescope of the Crimean Astrophysical Observatory. This array, which operates at wavelengths of 1.35 and 18 cm, could form the main ground-based array for a space radio interferometer. Experiments are carried out regularly with the participation of a number of organisations of the Soviet Union, USA, Sweden, the Federal Republic of Germany, Australia, England and Italy: at first two or three, and then four or five experiments per year.

The experiments conducted with the participation of IKI (Matveenko, L. R. Kogan, V. I. Kostenko and others) studied two types of objects: extragalactic objects with active nuclei (quasars, radio galaxies, BL Lac objects) and star-forming regions with H_2O and OH maser sources. Data were obtained about the detailed structures of quasars, the structure of their magnetic fields and their brightness temperatures, which proved to be 10^{12} K or higher for individual components in the source structure. In some cases, apparent motions of components faster than the speed of light were observed.

---

[2] At about the same time, work on the development of VLBI was begun at the Main Astronomical Observatory by N. L. Kaidanovskii (see Essay 4).

Simultaneous with this international program, IKI together with FIAN and the Crimean Astrophysical Observatory took part in establishing a Soviet interferometer operating between Simeiz and Pushchino. The specifications for this interferometer established by the Scientific Council on Radio Astronomy of the USSR Academy of Sciences and confirmed by the Division of Physics and Astronomy of the USSR Academy of Sciences foresaw the use of an interferometer comprised of the RT-22 radio telescopes of FIAN and the Crimean Astrophysical Observatory, together with a complex of computer-based equipment, hydrogen-maser frequency standards operating at 1.35 cm and other equipment. The construction of the interferometer was finished in the middle 1970s, and systematic observations of water-maser sources were begun in 1976. In 1982, the RT-70 telescope was added as another interferometer element. In addition, the new wavelength of 18 cm began to be used, which appreciably expanded the capabilities of the interferometer.

As a result of joint studies on this instrument and the global VLBI network at 1.35 cm, the structures of regions of star formation could be studied in detail. It was established that the sources of H_2O emission were concentrated in individual centres of activity about 1000 astronomical units in size, with the sizes of the sources themselves being about one astronomical unit. The maser sources in an activity centre were located in disks or rings, with a protostar with a mass of 1–20 times the mass of the sun. One set of important results obtained using the Simeiz–Pushchino interferometer were derived from studies of H_2O "flares" in the source Orion-4.

Investigations of Rapidly Variable Processes Using Synchronous Observations
Studies aimed at detecting radio flares arising during supernova explosions were undertaken at IKI at the beginning of the 1970s. All-directional antennas and wideband receivers operating at several decimetre and metre wavelengths were used. The receiving antennas were separated by large distances (thousands of kilometres), making it possible to exclude a large fraction of the local interference (by analysing impulsive events synchronous at different antennas). Signals originating in space could be identified from the delay in their signals at low frequencies relative to high frequencies; the magnitude of this delay, which is determined by the dispersion measure in the interstellar medium, could then, in principle, be used to estimate the distance to the source. This method became known as synchronous dispersional signal reception. Together with searches for supernova flares, this method was used to carry out searches for signals from extraterrestrial civilisations. This work was conducted by Popov, Soglasnov, E. E. Spagenberg and V. Sysoev under the direction of Kardashev. Workers at the Moscow State Pedagogical University, Sternberg Astronomical Institute and Moscow Energy Institute took part in the development of the equipment and in the observations. In 1972, observations were carried out at two points: in the Northern Caucasus and Pamir. The Caucasus station continued to operate in 1973, and a new station was set up at Kamchatka, in order to increase the baseline between the antennas, and thereby more effectively exclude interference from Earth-orbiting satellites. In addition, a French setup operating at metre wavelengths was installed at these ground stations and on board the *Mars-7* spacecraft in the framework of a joint Soviet–French experiment. These observations led to the

detection of several types of coincident impulses, due to radiation from the ionosphere (detected earlier by Gorkii radio astronomers), solar flares and interference from Earth-orbiting satellites. No cosmic signals with appreciable dispersions were detected. Synchronous searches of impulsive signals using widely separated omnidirectional antennas were ceased in the middle 1970s. The synchronous-dispersion method began to be used jointly with observations on directional antennas (the 22-m radio telescopes in the Simeiz and Pushchino and the 100-m radio telescope in Germany) in connection with investigations of individual peculiar sources.

In 1974, the first synchronous observations of pulsars aimed at measuring correlations of the fluctuations of the mean pulse intensities were carried out on the Simeiz and Pushchino 22-m telescopes. Unfortunately, it was not possible to obtain observations sufficient for the cross-correlation analysis due to the presence of interference in the Pushchino data. However, the Simeiz data yielded the mean profiles and scintillation curves for several pulsars (PSR 0329+54, PSR 0950+08, PSR 1133+16 and PSR 1929+10).

The experiment was repeated in 1977 with PSR 0329+54 at a frequency of 550 MHz (with a channel bandwidth 5 MHz), but the interference proved to be even worse. Therefore, it was not possible to obtain observations that were free from interference simultaneously at two different sites.

A more successful set of synchronous observations was conducted in 1977 and 1980 at Pushchino on the DKR-1000 telescope (102.5 MHz) and at Effelsberg (Federal Republic of Germany) on the 100-m telescope (1700 MHz), as part of a cooperative Soviet–West German programme with the participation of IKI and FIAN. The resulting data were used to determine the location of the radio-emitting region in PSR 1133+16. Correlations between the mean intensities of pulsar pulses at frequencies of 102.5–1700 MHz were also found, together with the frequency dependence of the drift parameters for PSR 0808+74. Finally, in 1979, an experiment was carried out synchronously between the 70-m radio telescope of the Deep-Space Communications Centre and the 10-m radio telescope on board the *Salyut-6* orbiting station.

During the processing of these two-antenna experiments at IKI, a data-registration complex based on an *Elektronika-100I* mini-computer and *IZOT* recorders was developed. This complex was also used for 102.5 MHz observations of pulsar pulses with high time resolution (to 10 microseconds) starting in 1976, conducted in collaboration with FIAN on the Large Scanning Antenna in Pushchino. A dispersion-compensation method was applied, making it possible to greatly reduce smearing of the pulses due to dispersion during the propagation of the radio waves in the interstellar medium. This method was used in observations of ten of the brightest pulsars, in order to derive the characteristics of the microstructure in their pulses.

The method of synchronous observations was also used in studies of the Galactic centre.

Submillimetre Astronomy This area began to be developed at IKI immediately after the formation of the Department of Astrophysics at the end of the 1960s, when observations on the 22-m Pushchino antenna were conducted jointly with GAISH (see above), and then continued in the Laboratory of Submillimetre Astronomy under Sholomitskii. An appreciable part of the laboratory's work was associated with the development of the main components of receiver nodes—band spectral filters, optico-mechanical modulators, etc. These components were tested in airplanes and on ground-based telescopes.

In 1979–1981, a series of observations of quasars was conducted on the 6-m optical telescope of the Special Astrophysical Observatory of the Academy of Sciences, in collaboration with the group of G. V. Schultz at the Max Planck Institute for Radio Astronomy in Bonn. Observations were obtained at 1 mm wavelength (270 GHz) with a sensitivity of 0.3–0.5 Jy for an acculation time of one hour. The fluxes of 20 quasars were measured with this sensitivity. Comparison with data obtained by other groups demonstrated that some of the quasars had their maximum flux at 1 mm.

Simultaneously, investigations into selecting a site for ground-based submillimetre observations were conducted. In 1980–1981, the water-vapour content and atmospheric opacity at Shorbulak (East Pamir), at an altitude of 4350 m, were analysed in collaboration with the Main Astronomical Observatory. These studies established that Shorbulak was a very high-quality site for ground-based submillimetre observations, which was not surpassed in terms of its atmospheric transparency even by the well known high-altitude observatory at Mauna Kea (Hawaii, USA, 4200 m altitude).

Theoretical studies. These were led at IKI by Shklovskii in the Department of Astrophysics. Although the scientific interests of Shklovskii made a substantial shift into the new and vigorously developing field of X-ray astronomy in the 1970s, he also continued to work intensively in radio astronomy. In 1974, he successfully applied the theory of a relativistically expanding synchrotron source developed by Rees to explain the jet in NGC 4486. Shklovskii treated this object like a succession of magnetised clouds of plasma—"plasmons"—ejected from the active nucleus of the galaxy at relativistic speeds. Because of relativistic Doppler beaming, the intensity of the radio emission of plasmons moving in the direction of an observer is sharply enhanced, while the emission in the opposite direction is suppressed, providing an explanation for the one-sided appearance of the jet. Shkovskii returned to the problem of one-sided jets in 1981, emphasising their importance for our understanding of the nature of quasars.

At the end of the 1970s and beginning of the 1980s, L. S. Marochnik (a student of S. B. Pikel'ner) and a group of theoreticians headed by I. D. Novikov joined the theoretical group in the IKI Department of Astrophysics. The main area of interest of this group was cosmology and relativistic astrophysics. However, a number of their studies had a direct connection to radio astronomy.

As early as 1968, Novikov showed that, if the Universe is uniform, the onset of the cosmological expansion was anisotropic and the density ρ was less than the critical density ρ_{cr}, the "relict radiation" should display a characteristic distribution on the sky: the intensity of this radiation should be nearly constant over the entire sky, except for a spot of enhanced (or reduced) intensity with an angular size of the order of $\Omega = \rho/\rho_{cr}$. These results were generalised to the case of arbitrary deviations from isotropy and spatially very large perturbations of the homogeneity of the Universe in studies by Doroshkevich, V. N. Lukash and Novikov.

Doroshkevich, Novikov and A. G. Polnarov examined the influence of cosmological gravitational waves on the observed anisotropy of the cosmic background radiation in 1977. In 1980, M. M. Basko and Polnarev carried out an exact calculation of the effect of the polarisation of the cosmic background radiation in the case of anisotropy in the expanding Universe. An important result of these calculations is that the degree of polarisation is very sensitive to heating of the intergalactic medium during the formation and evolution of galaxies.

Radio Astronomy and Cosmology. Radio Astronomy Studies in the IKI Department of Theoretical Astrophysics Starting in the middle of the 1960s, the extremely strong theoretical physicist Ya. B. Zel'dovich actively began a series of studies in astrophysics. Over a short time, he assembled a group of talented young scientists, many of whom came to his group directly from university. At first, Zel'dovich's group worked in the Institute of Applied Mathematics of the Academy of Sciences, also having close connections with GAISH, where the unified astrophysical seminars took place and where Zel'dovich gave a course for students in the Astronomy Department of Moscow State University. Early in 1974, part of the group was moved to IKI, where the Department of Theoretical Astrophysics was formed under the supervision of Zel'dovich. The main areas of interest of this department were associated with cosmology and relativistic astrophysics. Here, we will touch upon only those studies that had a direct connection to radio astronomy.

These were, first and foremost, related to the relict radiation. In 1966, immediately after the discovery of the cosmic background radiation, Zel'dovich gave an interpretation based on a hot Universe, and began to develop the theory of this hot Universe. A large series of works by Zel'dovich and his students were subsequently dedicated to this topic. In 1970, Zel'dovich and Sunyaev showed that, as a consequence of energy released in the early stages of the evolution of the Universe, the spectrum of the cosmic background radiation should differ from that for a blackbody. In a series of studies (1969, 1970), they analysed the variations of the spectrum expected for various energy losses. These effects were then studied in more detail by I. F. Illarionov and Sunyaev (1974). It was shown that there should exist two main types of spectral distortions arising in early stages of the evolution of the Universe. Thus, detailed analyses of the spectrum of the cosmic background radiation can yield the epochs for specific types of energy loss (effectively dating these losses).

A completely different type of distortion of the cosmic-background spectrum is due to recombination of hydrogen in the Universe at a redshift of $z \approx 1500$. As was demonstrated by Zel'dovich, Kurt and Sunyaev in 1968, radiation associated with two-photon transitions and Lyman-α emission lead to the appearance of bands in the submillimitre part of the spectrum. Calculations of the recombination for a non-equilibrium cosmic-background spectrum carried out in 1983 by Yu. E. Lyubarskii and Sunyaev showed that the recombination lines of hydrogen and helium were accessible to observations using radio telescopes with the sensitivity available at that time.

A number of papers by Zel'dovich and Sunyaev were dedicated to studies of fluctuations of the cosmic background radiation. In 1970, they calculated the fluctuations arising at the epoch of recombination, $z \approx 1500$, due to scattering of the background radiation on electrons moving in a field with only small density inhomogeneities, whose subsequent growth led to the formation of galaxies and clusters of galaxies. Thus, it was shown that studies of the fluctuations of the cosmic background radiation could provide very valuable information about the formation of galaxies and galaxy clusters.

The observed background fluctuations are due to both genuine fluctuations in the cosmic background radiation and fluctuations in the distribution of radio sources. Their contribution was studied by M. S. Longair (England) and Sunyaev in 1968. More detailed investigations of the fluctuations $\Delta T/T$ on various angular scales θ were carried out by P. B. Partridge (USA) and Sunyaev in 1982, whose calculations indicated a minimum in the function $\Delta T/T(\theta)$ on an angular scale of $\theta \sim 10'$. This determines the most promising scale for searches for intrinsic fluctuations.

The next series of works was associated with counts of radio sources. Observations show that the growth in the number of sources with increasing redshift z ceases at $z = 2$. This led to the idea that all radio sources were born at $z < 2$. In 1968, Doroshkevich, Longair and Zel'dovich established that another completely different interpretation was possible: radio sources were born at $z > 2$, but live a finite time, over which there was a "decay", or sharp decrease in the number of sources (analogous to radioactive decay) according to an exponential law. This model to explain the "evolutionary effect" is currently widely accepted.

An important research direction was associated with studies of clusters of galaxies. In 1973, Zel'dovich and Sunyaev determined that, during the formation of clusters, in the evolution of so-called pancakes (protoclusters of galaxies), there is a stage ($z \sim 3\text{--}10$) when their central regions are in the form of gigantic ($M \approx 10^{13}\text{--}10^{14}$ solar masses) clouds of neutral hydrogen at a temperature of $T \sim 10^4$ K. Such clouds could be observed via their 21-cm lines, shifted toward the red to metre wavelengths.

In 1972–1982, Zel'dovich and Sunyaev carried out a series of investigations into interactions of the cosmic background radiation and the intergalactic gas in clusters of galaxies. They showed that, when background photons are scattered on electrons in the hot intergalactic gas, they are shifted toward higher frequencies, making clusters powerful sources of submillimetre and short-millimetre wavelength radiation. The opposite effect is observed at centimetre and longer millimetre ($\lambda > 2$ mm) wavelengths: the cluster becomes a negative (!) source; in other words, scanning of the sky should reveal a reduction in the brightness of the cosmic background radiation in the direction toward a cluster. This so-called Sunyaev–Zel'dovich effect was observed in a number of galaxy clusters by Y. N. Pariiskii in 1972, as well as by English, American and West German radio astronomers. Combined with the information from X-ray observations, measurements of this effect can be used to determine the linear size of the cluster, and therefore its distance and the Hubble constant. Another important conclusion that flows from the existence of the Sunyaev–Zel'dovich effect is that it provides experimental proof that the cosmic background radiation was created in early stages of the evolution of the Universe.

Zel'dovich and Sunyaev obtained a fundamentally important result in 1980: they showed that all clusters of galaxies could be "tied" to an absolute coordinate system defined by the condition that the cosmic background radiation is isotropic in this system. Consequently, the cosmic background radiation is playing the role of a new "ether". If a galaxy cluster with gas moves relative to this coordinate system with a speed v, the brightness of the cosmic background radiation in the direction of the cluster will vary in proportion to v/c, and a polarised component of the radiation will appear. Measurements of the brightness and polarisation of the cosmic background radiation in the cluster direction can be used to derive the velocity of motion of the cluster relative to the background radiation.

Other research that forms part of the series of studies concerned with clusters of galaxies is Sunyaev's work on their diffuse radio emission. He showed that, if there is a bright source in the cluster, scattering of its radiation on the intergalactic gas leads to a diffuse radiation component in the cluster, characterised by an extremely high degree of polarisation, $\sim$33–66%. This effect was calculated for the Virgo and Per A clusters, and observed with the RATAN-600 telescope and the 100-m Effelsberg radio telescope. Since the diffuse radiation is delayed relative to the direct radiation from the source, it provides information about the brightness of the source at earlier epochs (up to a million years earlier than the epoch of the direct radiation from the source).

Another series of studies by Sunyaev was concerned with observations of radio lines. In 1966, he noted the importance of observations of hyperfine splitting of the $\lambda = 3.46$ cm line of the He^{3+} ion, and calculated the parameters of this line, which was detected in observations made with the 100-m Effelsberg telescope. In 1968, L. A. Vainstein and Sunyaev performed calculations of two-electron recombination, and showed that it can lead to extremely strong population of the upper levels of a number of ions. Lines arising due to transitions between these levels should be observed in the spectra of quasars and of the Sun at centimetre to millimetre wavelengths, however, these lines have not been detected thus far.

In 1969, Sunyaev pointed out that observations of the periphery of a galaxy in the 21-cm line could provide unique information about the spectrum and intensity of the background ionising radiation of the Universe ($\lambda < 912$ Å). The ionisation of hydrogen at the edge of the galaxy by this radiation should be reflected in the 21-cm isophotes. Since this radiation is absorbed by neutral hydrogen at the periphery of our own Galaxy and therefore does not reach the solar system, this provides the only possibility of using 21-cm radio observations to estimate the intensity of the background ionising radiation.

The 1972 work of V. S. Strel'nitskii and Sunyaev on the acceleration of H_2O maser sources near hot O stars is of considerable interest. The acceleration is brought about by the radiation of the hot stars and by shocks. This can explain the high velocities (up to 200 km/s) with which H_2O maser sources leave the star-formation regions with which they are associated for the interstellar medium (the "Flying Dutchman" effect).

The works of Zel'dovich and his students laid new pathways in radio astronomy, closely associated with cosmology. This research transformed radio astronomy from a science concerned with various individual processes in the Universe, into a science providing valuable information about the evolution of the Universe as a whole.

3.6 Radio Astronomy at GAISH in the Post-transition Period

Radio astronomy research at GAISH developed in the following directions during the 1970s: surveys of the sky on the RATAN-600 telescope, spectral studies, studies of kilometre-wavelength cosmic radio emission using instruments on board spacecraft, submillimetre astronomy and theoretical studies.

Sky Surveys on the RATAN-600 In accordance with the ideas behind its design and realised by its construction, this radio telescope is excellently suited for all-sky surveys. The knife antenna beam makes it possible to carry out comparatively rapid surveys of an appreciable area of the celestial sphere at centimetre wavelengths with high angular resolution in one coordinate. Another important feature is the ability to conduct observations simultaneously at several frequencies. This makes it possible to obtain instantaneous radio spectra during a survey. In contrast to earlier surveys, a multi-frequency survey on the RATAN-600 at 2, 3.5, 7.6 and 18 cm is being planned. Several receiver complexes operating at these wavelengths are being constructed in the RATAN-600 laboratory (partially on its own, partially in collaboration with other organisations). This work was done under the supervision of V. R. Amirkhanyan. The main wavelength of the survey is 7.6 cm.

Large databases require fully automated observation and data-processing procedures. With this idea in mind, a digital and computational complex was developed at GAISH under the supervision of M. G. Larionov. The main principles behind the mathematical reduction algorithms were due to A. G. Gorshkov, and work on the mathematical side of the complex was carried out under his supervision.

The survey observations were begun in 1978. The survey was carried out on the Southern sector of the RATAN-600 using two radiometers at 2 and 3.5 cm. A radiometer operating at 7.6 cm was added in 1979.

A survey of a section of the sky at declinations 0–7.5° was conducted in 1981, with the participation of Amirkhanyan, Gorshkov, A. A. Kapustkin, V. K. Konnikova, A. N. Lazutkin, Larionov, A. S. Nikanorov, V. N. Sidorenkov, L. S. Ugol'kova and O. N. Khromov. Of 650 sources detected in a 0.17-steradian area of the sky, 370 proved to be new objects that had never been mapped before; in other words, the number of new sources detected in the survey exceeded the number of previously known sources. This fully confirmed preliminary estimates made by Kardashev in 1964, who argued for the importance of carrying out surveys at centimetre wavelengths. This was the first survey in the Soviet Union that yielded a large number of new radio sources (apart from the Khar'kov survey, conducted at decametre wavelengths).

Analysis of the instantaneous spectra of the radio sources found in this survey showed that the mean spectral index at 2–3.5 cm was 0.4 ± 0.4, roughly the same as had been obtained for non-instantaneous spectra at centimetre wavelengths. This was a very important result, since it indicated that the spectral characteristics of variable radio sources could be studied using non-simultaneous observations at different wavelengths.

The distribution of the 7.6-cm survey sources was found to be uniform on the sky. The survey data provided information about the angular (and linear) sizes of an appreciable number of quasars and galactic nuclei, and confirmed the dependence of the number of observed sources on their radio flux derived earlier using survey data at other frequencies, which indicates the presence of substantial evolutionary effects.

Spectral Observations Productive studies of OH radio lines were carried out at GAISH in the 1970s using the Nancay radio telescope (Lekht, Pashchenko, Rudnitskii). Observations at 18 cm required a number of modernisations of the telescope, and a new receiver system was developed by staff of GAISH in collaboration with colleagues at the Meudon Observatory.

Interest in OH maser sources was stimulated by the role they played in star formation (the idea that there was a connection between maser sources and the early stages of star formation was first put forth by Shklovskii). In 1972, Pashchenko and Slysh discovered dense molecular clouds 3–5 pc in size with densities of 10^3 cm^{-3} near type I OH maser sources, via their absorption in the OH lines. A physical relationship was established between these clouds, the sources of type I OH masers and compact HII regions. Several years later, these same researchers found regions of extended maser emission in satellite lines near compact type I OH masers that were radiating intensely in the main OH lines. It was shown that this extended maser emission comes from the dense molecular clouds discovered earlier. This opened the possibility of obtaining information about the structure and physical parameters of the molecular clouds from observations in the satellite OH lines (together with observations of their absorption in the main lines). This essentially represented a new method for studying the interstellar medium, which is especially effective in (the many) cases when there are no regions of extended continuum emission in the studied region that could be used to measure absorption lines. In addition to studying known sources, searches for new OH maser sources were carried out, leading to the discovery of 22 new sources.

Studies of maser sources emitting in water-vapour lines at 1.35 cm in active star-forming regions were begun in 1979. A 96-channel spectral analyser with high spectral resolution was constructed at GAISH for this purpose. Observations of star-forming regions in different molecular lines (OH and H_2O) enable the investigation of different stages in the star-formation process. The 1.35-cm observations were carried out using the 22-m FIAN radio telescope in Pushchino. The variability of more than 60 H_2O maser sources was studied in collaboration with R. L. Sorochenko of FIAN. Together with sources related to active star-forming regions were maser sources associated with variable infrared stars (giants and supergiants), which were in a late stage of stellar evolution.

In addition to their studies of maser sources, the spectral group of GAISH (Lekht, Pashchenko, Rudnitskii) carried out searches for radio recombination lines of hydrogen and carbon at metre wavelengths ($\lambda = 3$ m). Observations of Sagittarius A,

Omega, W 51 (searches for hydrogen lines) and Cassiopeia A (searches for carbon lines) were conducted using the DKR-1000 and Large Scanning Antenna of FIAN. These observations provided upper limits for this line emission that enabled refinement of the parameters of these sources and the construction of models for gaseous clouds they might contain.

Studies of Long-Wavelength Radio Emission The research begun by Kardashev and Slysh in the 1960s was successfully continued in the 1970s under the supervision of V. P. Grigorieva. These studies were carried out at 50–2000 kHz, and were aimed primarily at investigating type II and type III solar radio outbursts, which arise in the solar corona and in interplanetary space during the propagation of shock waves and streams of electrons with energies of 10–1000 keV. Since the radio emission associated with these processes drifts in frequency, receivers with several frequency channels are required for such studies.

A multi-channel device developed at GAISH was installed on satellites of the *Prognoz* (*Prognoz-1*, *2*, *3*, *4*, *5*, *8*) and *Luna* (*Luna-11*, *12*, *22*) series. Observations of type-II outbursts were used to reconstruct the distribution of shock-wave velocities in the corona and interplanetary medium at distances from the Sun of 10 solar radii to the Earth's orbit. The kinetic temperature of the gas in these shocks and the brightness temperature of the radio emission from regions of plasma turbulence were derived. A correlation was found between the brightness temperature and velocity of the shock waves. Observations of type-II outbursts over many years were used to compare the parameters of the associated shock waves at different phases of the solar activity—the first time such a comparison was done. The distribution of electron velocities was derived from observations of type-III outbursts. All these studies provided input for models of the distribution of the electron density in the corona and interplanetary medium.

Moving into the Millimetre and Submillimetre Bands Zabolotnii at GAISH continued to work on pushing into these wavebands in the 1970s. In 1972–1975, he and V. F. Vystavkin (Institute of Radio Physics and Electronics) developed a high-sensitivity, broadband millimetre radiometer whose operation was based on the Josephson effect. Observations using this radiometer were carried out on the 22-m radio telescope of the Crimean Astrophysical Observatory in 1976. The first submillimetre (at wavelengths of 350 and 450 micron) observations in the Soviet Union were conducted in 1975, at the High Mountain Station of GAISH at Zailiiskii Alatau, at an elevation of about 3000 m above sea level. The opacity of the atmosphere at 350–450 micron was measured, and the possibility of conducting successful ground-based astronomy observations at these wavelengths under high-mountain conditions demonstrated. Unfortunately, these studies did not receive the necessary support from GAISH, and this line of research had to be ceased, with the result that Zabolotnii moved to IKI.

Theoretical Studies Theoretical research in radio astronomy at GAISH was carried out by Shklovskii, Pikel'ner and Kuril'chik. The results obtained by Shklovskii in the 1970s are described above.

Pikel'ner's studies associated with the interpretation of observational data for small-scale inhomogeneities in gaseous nebulae stand out as being among his most important work from this period. In 1973, he and Sorochenko (FIAN) deduced (based on optical and radio data) the presence of numerous small clumps of gas with electron densities of 10^4 cm^{-3} in the Orion Nebula. Pikel'ner suggested that these clumps originated due to the compression of initial fluctuations by shock waves arising during the deceleration of stellar winds.

Beginning in the 1960s, Pikel'ner was very interested in questions related to solar magnetohydrodynamics, and, more generally, a wide spectrum of problems in solar physics. His fundamental ideas in this area were generalised in the 1977 book *Physics of the Plasma of the Solar Atmosphere*, which he wrote in collaboration with Kaplan and V. N. Tsytovich. It is interesting that his studies of type-II solar radio outbursts led Pikel'ner to broader investigations of the interactions of shock waves in plasma, which he studied intensively together with Tsytovich in the 1960s and 1970s, as applied to both solar physics and the interstellar medium.

Kuril'chik carried out theoretical research in two areas: (1) statistical studies of the observed parameters of extragalactic sources and (2) the development of models for radio sources. The models he developed describe various structures that can be distinguished in radio sources and the phenomena occurring in them. For example, he constructed a model for the circum-nuclear structures responsible for the variable and "background" (quasi-stationary) components of the centimetre-wavelength radio emission. He also constructed a model for the structures displaying the characteristic variability observed at decimetre wavelengths, as well as other models. Kuril'chik's new interpretation of the jet in the radio galaxy Virgo A is of considerable interest. In this interpretation, based first and foremost on polarisation data, the optical and radio knots are due to anisotropic radiation from flows of particles in loops of magnetic field within the overall bipolar configuration.

This chapter could not be written without the help of colleagues and staff of the Sternberg Astronomical Institute (GAISH) and the Space Research Institute (IKI), who presented material, read over some or all of the manuscript and made valuable comments. I am sincerely grateful to all of them. I am especially thankful to N. G. Bochkarev, A. G. Gorshkov, V. P. Grigorieva, V. F. Zabolotnii, N. S. Kardashev, V. N. Kuril'chik, E. E. Lekht, L. I. Matveenko, I. D. Novikov, M. V. Popov, V. I. Slysh, R. A. Sunyaev, I. S. Shklovskii, G. B. Sholomitskii and G. S. Tsarevskii.

Chapter 4
The Department of Radio Astronomy of the Main Astronomical Observatory of the USSR Academy of Sciences

N.L. Kaidanovskii

Abstract The history of the organisation of the Department and the principles behind the construction of unique telescopes in the form of variable-profile antennas developed at the Main Astronomical Observatory (GAO) are described, together with the development of other new methods and equipment and astrophysical results obtained by the GAO radio-astronomy group. The chapter also discusses the birth of the RATAN-600 radio telescope and the beginning of work on its realisation.

The Department of Radio Astronomy of the Main Astronomical Observatory of the USSR Academy of Sciences in Pulkovo (GAO) was born in the Lebedev Physical Institute (FIAN), organised by the Head of Radio Astronomy Section of the Oscillation Laboratory of FIAN S.E. Khaikin.

The principles behind the construction of variable-profile antennas (VPAs) were developed at FIAN in 1952, and the construction of the Large Pulkovo Radio Telescope, intended to become the main instrument of the GAO Department of Radio Astronomy was planned. Work on the perfection of radiometers and polarimeters operating at centimetre wavelengths was carried out at FIAN.

4.1 Choosing a Research Direction. Principles Behind the Construction of Variable-Profile Radio Telescopes

When determining the research programme for the new Department, Khaikin proceeded from the following basic ideas laid out by him earlier in the collection *The Pulkovo Observatory's 125th Anniversary*.[1]

The sensitivity and resolution that can be attained in radio astronomy is limited by the size of the antennas and sensitivity of the receivers used. These limitations depend on the observing wavelength within the "window of transparency" of the Earth's atmosphere. Early in the development of radio astronomy (approximately up to 1954), technical difficulties were easiest to overcome at metre wavelengths,

---

[1] A.A. Mikhailov (ed.), *125 years of the Pulkovo Observatory*, 1966, Leningrad: Nauka (in Russian).

where it was possible to construct and use antennas and radio interferometers with very large areas and relatively inaccurate surfaces, since the allowed errors could be as large as several tens of centimetres. The receivers had lower intrinsic noise levels than receivers operating at decimetre and centimetre wavelengths, making them more sensitive. Metre wavelengths are also favourable for observations of the most numerous cosmic radio sources: sources of non-thermal radiation, whose spectral fluxes grow with increasing wavelength. All these factors facilitated in this initial period the development of observational radio astronomy primarily at metre wavelengths. It was also thought that it would be possible to achieve the best penetrating power at these wavelengths. In those distant years, the centimetre and short-decimetre wavelengths at which the thermal emission of Galactic nebulae, planets and the neutral-hydrogen line were studied played a less important role. However, Khaikin understood the error in this attitude, and energetically insisted on moving radio astronomy to centimetre wavelengths, considering this range to be most promising for essentially all radio astronomy studies. In this range, the sky noise due to the radiation of the Galaxy and terrestrial atmosphere are minimum, and the receiver noise temperatures can be lowered to values that are close to absolute zero, which he argued should compensate for the decrease in the spectral fluxes of non-thermal sources with decreasing wavelength. It follows that all sources that can be observed at metre wavelengths should also be accessible to observations at centimetre and short-decimetre wavelengths, as well as sources that cannot be observed at metre wavelengths.[2] Therefore, centimetre wavelengths played the main role in the research programme Khaikin created for the Pulkovo Observatory Department of Radio Astronomy.

Nowadays, when the noise temperatures of receivers with parametric or paramagnetic amplifiers cooled to liquid-helium temperatures can achieve values as low as a few Kelvin (and this is now quite standard), the predictions of Khaikin with regard to the centimetre-wavelength band seem trivial. However, we should not forget that his views were expressed at a time when the best centimetre-wavelength receivers had noise temperatures of 4000–5000 K, and it was possible to barely detect only a few of the brightest supernovae or thermal nebulae with telescopes operating in this waveband, due to their low sensitivity, the instability of the amplification and interference due to reflections of noise from the receiver input elements from the feed.

We must also credit the persistence of Khaikin, who defended his fundamental positions when they were subject to energetic attacks from radio astronomers who were proponents of giving preference to metre wavelengths.

In order to substantially progress radio-astronomy studies at centimetre wavelengths, in addition to sensitive receivers, antennas with high resolutions and large receiver areas were needed. At that time, the principle of consistent (homologous) deformations was not yet known, and it was not possible to achieve a relative accuracy in the construction and preservation of the surface shape, determined by technological and deformational deviations, of better than 10^{-4}, even in the best

---

[2] Pulsars with very steep spectra whose detection is inefficient at centimetre wavelengths were not known at that time.

parabolic reflectors. Since deviations from the specified surface shape should not exceed one-tenth of a wavelength, a relative accuracy of 10^{-4} implies that the diameter of a reflector cannot exceed 1000 wavelengths. With such a reflector diameter, the opening angle of the antenna beam will be no smaller than $3'$. Further increase in the resolution required new techniques enabling the construction of much larger reflectors with relative accuracies much higher than 10^{-4}.

Such a new technique was proposed by Khaikin and N. L. Kaidanovskii, and used as the basis for the development of what was initially called a "fan antenna," and subsequently acquired the name of a "variable-profile antenna" (VPA). In view of its importance for the discussions below and the non-trivial nature of this principle, we will first consider it in more detail, together with the constructional realisation of such an antenna.

The proposed technique was that the reflecting surface of an antenna be composed of a large number of identical, mechanically unconnected reflecting elements, which are mounted so that they together form the required shape for the reflecting surface. As the size of the reflector is increased, due to the increase in the number of reflecting elements, it is the relative accuracy of their positioning that must be enhanced, rather than the relative accuracy of each individual element. The problem of adjusting the reflector is solved using geodesic and remote phase-measurement methods, which make it possible to achieve a relative accuracy better than 10^{-6}. This is an order of magnitude better than the accuracy that can be reached in a non-multi-element reflector, even one constructed based on the principle of homologous deformations, such as the 100-m Effelsberg radio telescope (Germany, with a relative error of 10^{-5}), making it possible to obtain much higher resolutions for the same telescope area.

The reflecting elements of a VPA are placed near the Earth's surface in a horizontal orientation. Since the horizontal size of the antenna is appreciably larger than its vertical size, its beam has the form of a vertical knife (a knife beam).

A VPA must have the ability to establish the vertical knife beam in various azimuths. Due to the Earth's daily rotation, the knife beam will intersect an observed source in various directions, which, in principle, provides the possibility of obtaining the same information provided by observations with a needle-like beam whose opening angle is equal to the minimum opening angle of the knife beam.

The main reflector of a VPA transforms a plane wave from a distant source to a cylindrical wave that converges on the focal vertical, where it can be received by either a linear antenna (for example, a feed array) or a secondary reflector that transforms the cylindrical wave to a spherical wave that converges onto the phase centre of the primary feed. This secondary reflector can be a parabolic cylinder with a horizontal element.

The surface of the main reflector that is directed towards a source at an hour angle of h forms a section of an elliptical cone with its axis inclined at an angle of $\pi/4 - h/2$ to the vertical, contained between two horizontal planes. The sections of the cone by these horizontal planes form ellipses with the same eccentricity, $e = \cos h$. The foci of the ellipses lie on the vertical line passing through the apex of the

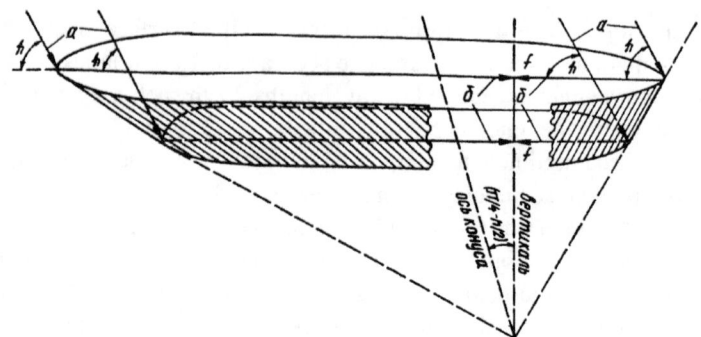

Fig. 4.1 Schematic of the shape of the VPA reflecting surface, corresponding to a horizontal cut of an elliptical cone with its axis inclined to the vertical by $\pi/4 - h/2i$ (this angle is labelled; the Russian *text on the right and left* reads "vertical" and "cone angle"): (a) incident rays; (b) reflected rays; (f–f) focal line; (h) angle of incidence of the rays

cone (Fig. 4.1). The cylindrical wave reflected by the main reflector is focused on this vertical line.

Within the illumination angle, the profiles of the ellipses in the sections are close to circular, and can be constructed by radially shifting the elements of circles and rotating them in the azimuthal direction. Since a VPA should be able to provide observations at any azimuth and at any hour angle from $h = 0°$ to $h = 90°$, the reflecting elements are given a cylindrical rather than a conical shape; the radius of curvature and size of the reflecting elements in the transverse and longitudinal directions must be limited so that the deviation of the reflector elements from the theoretical conical surface does not exceed 1/20 of a wavelength.

To form a reflecting surface appropriate for the coordinates of a source, the reflecting elements must be outfitted with some mechanism to bring about progressive radial shifts and rotations in azimuth and hour angle, and must be mounted on a closed circular foundation with a radius R.

In order to minimise the required radial shifting, an optimal initial cone is chosen for each hour angle. This makes it possible to decrease the maximum radial shift ΔT by approximately a factor of ten, and to reduce the value of $\Delta R/R$ to 0.3%.

The antenna feed is placed on a trolley, making it possible to mount it at the antenna axis for any azimuth at a distance from the reflector that corresponds to the elevation of the source.

The main principles behind and the possible constructional realisation of a VPA were written up as a proposed invention. Khaikin and Kaidanovskii obtained recognition as the authors of this invention starting from April 1, 1954. In the proposal, the technical basis for the new principle for the construction of the reflecting optics, which enabled expansion of the aperture size beyond the limits due to all forms of deformation, was also first formulated in this proposal.

4.2 Organisation of the GAO Department of Radio Astronomy

In a presentation to the Scientific Council of the Main Astronomical Observatory on May 30, 1952, Khaikin defined the task of the new Department of Radio Astronomy to be the measurement of the coordinates, spectral intensities and polarisations of cosmic sources, and the study of the monochromatic radiation of neutral hydrogen with high spectral and spatial resolution. The promise offered by improving radio astronomy methods was also noted. However, Khaikin did not only formulate these tasks, he also proposed realistic plans for their realisation. He intended the Department of Radio Astronomy to become one of the largest departments in the observatory. For him, it was essential to construct a separate two-story building and a number of radio telescopes operating at centimetre and decimetre wavelengths, for both astrophysical and applied purposes. The main instrument of the Department was to be the Large Pulkovo Radio Telescope (LPRT)—a variable-profile antenna with a diameter of 200 m. However, the reflecting elements would first fill only a northern quarter of the circle, whose chord had a length of 130 m. As a consequence, observations on the LPRT were possible only in the south at angular distances of $\pm 15°$ from the meridian. It was proposed that the LPRT would consist of 90 flat elements 1.5 m × 3 m in size.

In addition, several fully steerable 3-m and 4-m diameter antennas were planned. All the telescopes were to be outfitted with modern, sensitive receivers.

The organisation of the GAO Department of Radio Astronomy had already begun in 1952, but the Department was officially opened only at the beginning of 1954. Professor Khaikin became the Head of Department.

For the first radio astronomy observations, two 4-m diameter radio telescopes constructed using the design of Kalachev were transfered from FIAN to GAO in 1955. These were equipped with a polarimeter and radiometer designed to operate at 3 cm wavelength at low effective temperatures. The polarimeter was developed by E. G. Mirzabekyan and the radiometer by T. A. Shmaonov, both PhD students at FIAN; the latter came to Pulkovo for the final installation of the apparatus and the observations themselves.

The construction of the LPRT is an instructive tale. In 1953, Khaikin, having found a means to quickly manufacture inexpensive rigid girders of the LPRT reflecting elements from cast iron, instructed the FIAN construction engineer Kalachev to develop the moulds for the casting. A draft version of an LPRT reflecting element made of cast iron was prepared within a few days. Soon, Khaikin had obtained an agreement for the Syzranskii Heavy Machinery Plant to prepare 90 reflecting elements together with their cast-iron girders, which required further mechanical work. At the beginning of 1955, the first experimental example of a reflecting element was accepted at the plant by Kaidanovskii, representing GAO. All 90 reflecting elements for the LPRT were delivered to the Department of Radio Astronomy in Pulkovo by the end of 1955. Unfortunately, the construction of the building and foundation for the LPRT proceeded at a slow pace. The building for the Department was finished only in September 1956, and the construction of the LPRT was completed on November 12, 1956.

Fig. 4.2 Professor S. E. Khaikin with the first staff of the Division of Radio Astronomy of the Main Astronomical Observatory of the USSR Academy of Sciences (1954). *Left to right*: D. V. Korol'kov, T. M. Egorova, N. F. Ryzhkov, S. E. Khaikin, A. B. Berlin, V. N., Ikhsanova, K. P. Butuzov

As the LPRT reflecting elements were assembled, a preliminary adjustment was performed using geodesic methods. The first movable segmented parabolic feed designed for operation at 3 cm was built in the Department workshop. The adjustment and mounting of the instrument was begun.

Still two years before the official opening of the GAO Department of Radio Astronomy, two young specialists were transfered from FIAN—N. F. Ryzhkov and T. M. Egorova. They were responsible for the implementation of the new radio instruments and the preparation of a radiometer for spectral observations of hydrogen and deuterium lines. Their primary assistant was N. S. Efgrafov. GAO took on its first PhD student in 1953—the astrophysics graduate of Leningrad State University V. N. Ikhsanova.

In the beginning of 1954, in addition to the employees indicated above, two graduates of the Leningrad Polytechnical Institute were taken on by the Department— the radio engineers D. V. Korol'kov and K. P. Butuzov, as well as the radio technician A. B. Berlin as their assistant (Fig. 4.2). In 1955, the senior scientist A. P. Molchanov moved to GAO from Leningrad State University together with a group of radio physicists (N. G. Peterova, V. F. Kushnir, G. S. Veisig). This group was occupied with studies of the radio emission of the Sun. The senior scientist Kaidanovskii was also moved from FIAN to GAO. A number of astrophysicists moved from Moscow—the graduates of Moscow State University and students of Shklovskii Yu. N. Pariiskii (who became a PhD student in the Department) and N. S. Soboleva (taken on as a junior scientist). The construction engineers O. N. Shivris and A. I. Kopylov were taken on in that same year, as well as the graduates of the Leningrad Polytechnical Institute A. F. Dravskikh, G. P. Apushkinskii, G. M. Timofeeva and many others. The personnel of the Department continued to be rapidly augmented with young specialists in subsequent years. Astrophysicists began to work in the Department: G. B. Gel'freikh from Leningrad State University and the student of Shklovskii I. V. Gosachinskii. In accordance with a collaborative agreement, the Radio Physics Section of Leningrad Polytechnical Institute took on the calculation of the electrical characteristics of the LPRT and participated in the observations. This work was conducted by the Doctor of Technical Sciences B. V. Braude and the PhD student N. A. Esepkina, whose fruitful participation in the working of the Department of Radio Astronomy continues to the present day. The PhD in technical sciences A. A. Stotskii joined the Department of Radio Astronomy in 1961.

The selection of the Department staff was unusually successful. With only a few exceptions, all of the workers making up the initial group are still there to this day. This group took on the full task of forming the Department, constructing the antenna and developing extremely complex instruments and various observing methods, and never shunned "dirty work;" when necessary, the engineers and astrophysicists rolled up their sleeves, took up shovels and dug the pit for the foundation of the antenna, participated in the assembly of the LPRT, etc.

The maximum population of the Department of Radio Astronomy reached 80 persons, in 1965. Of the pioneers of the Department, one became a corresponding member of the USSR Academy of Sciences, seven became Doctors of Science and 16 became PhDs.

Young specialists of various Soviet republics—Armenia, Azerbaijan, Turkmenistan, Latvia—obtained training in radio astronomy and defended PhD theses in the Department. The Department helped in the training of Ukrainian and Siberian radio astronomers as well.

The scientific growth of the group, as well as its creative activity and independence were facilitated by Khaikin's style of supervision—a keenness for novelty and a sound estimation of fresh ideas, encouragement of initiatives by young coworkers and confidence in these young scientists.

Khaikin was also interested in the scientific problems of radio astronomy techniques and observing methods, and this work occupied an important place in the plans of the Department.

4.3 Development of Radio Astronomy Techniques and Methods

The first investigations by young GAO radio astronomers in 1953–1954 were a continuation and refinement of traditional studies carried out at FIAN.

Korol'kov created a stabiliser for the radio-lamp filament using magnetic amplifiers, which yielded a stability of $(3–5) \times 10^{-4}$. Under the supervision of Khaikin, Egorova and Korol'kov developed an original device for integrating weak signals with a time constant of the order of several hours. This was based on the use of a vibrational galvanometer, a synchronous detector in the form of a bridge with a photoresistor and a mirror galvanometer that wrote the detected deviations onto a photoplate. This device made it possible to distinguish a signal with a signal/noise ratio of 10^{-5}, and was used by Egorova and Ryzhkov in 1955 in their attempt to detect the 91.2-cm line of excited hydrogen.

The next important accomplishment of the Department was Korol'kov's modernisation of the polarimeter designed by Mirzabekyan at FIAN and then given to GAO. This polarimeter had a serious deficiency—appreciable parasitic linear polarisation associated with the difference in the beams in the E and H planes. Korol'kov proposed and realised a counter-reflector for the primary feed with a radial structure, which acted to symmetrise the beam and made it possible to decrease the parasitic polarisation to 0.1%. In addition, a ferritic isolator was placed in the high-frequency tract.

In the Summer of 1956, Korol'kov and Soboleva began observations with this polarimeter. This device subsequently served as an example for many polarisation-sensitive solar radio telescopes operating in various wavelength bands. The polarimeter proved to be especially fruitful for observations during solar eclipses. The high resolution provided by eclipse observations ($2''-3''$) enables measurement of the fine structures, spectra and polarisations of radio-emitting regions above solar spots, opening possibilities for measuring magnetic fields in the corona.

In 1956–1957, the construction engineer Kopylov designed and, with the help of plants in Leningrad, built a small series of fully steerable radio telescopes with equitorial mounts and with diameters of 2 and 3 m. These were outfitted with polarimeters and installed at Pulkovo and the GAO Kislovodsk Mountain Station, and later, starting in 1969, in Cuba. They were used during expeditions in connection with observations of solar eclipses (Fig. 4.3).

Gel'freikh and Korol'kov designed a small-baseline radio interferometer (SBRI) especially for observations of the solar radio emission during eclipses. The use of this phase-modulation interferometer made it possible to attenuate the signal from the "quiescent" Sun by more than a factor of ten during eclipse observations, thereby enhancing the accuracy with which the radio fluxes, coordinates and brightness distributions of local sources could be determined. This interferometer was also used to detect weak flares in the solar radio emission.

A. P. Molchanov developed and applied a scanning method for solar observations at centimetre wavelengths. Measurements were made of the shift in the effective centre of the solar radio emission, leading to the suggestion that a special solar radio-emission service be initiated, which is still operating in the Soviet Union.

However, the main direction in the methodical work of the Department was the realisation of and studies using the Large Pulkovo Radio Telescope. While the construction and assembly of the LPRT was underway, intensive work unfolded in the Department, on the improvement of this instrument, calculation of its electrical and geometrical characteristics, searches for methods for carrying out adjustments and antenna measurements, preparation of the programme of future radio astronomy observations and the methods to be used for these observations (Esepkina, Shivris, Pariiskii). Important ideas making it possible to enhance the sensitivity and resolution of an instrument of this type were put forward and realised. Therefore, although the development of the main principles behind an instrument comprised of reflecting elements was conducted at the Lebedev Physical Institute, the LPRT is properly

Fig. 4.3 Three-metre GAO antenna with an equatorial mount installed in Cuba

considered the work of the GAO Department of Radio Astronomy. Many members of the Department made substantial contributions to its realisation.

The leader of the group, Khaikin, proposed the main principles of the variable-profile antenna and supervised the development and realisation of the project.

Kaidanovskii carried out the necessary optical and geometrical calculations for the VPA, proposed the main principles for its construction, and searched for ways to appreciably shorten the movements of the reflecting elements, which cardinally simplified the building of the antenna, and supervised the construction and the first radio astronomy observations.

Esepkina carried out all the radio-technical calculations for the VPA, and proposed an original method for studying the antenna using a generator located in the near zone (without this, it would have been impossible to test the antenna using ground-based equipment). Further, this method was used to experimentally study the directional beam of the LPRT and the influence of errors in the mounting of the reflecting elements and feed.

Pariiskii and Khaikin showed that, for the vast majority of problems in radio astronomy, the resolution of a radio telescope was no less important than the need for high sensitivity. This makes it worthwhile to construct radio telescopes with vertical knife beams that could be used to observe at any azimuth, since observations with such telescopes could realise azimuthal aperture synthesis. They elucidated that the use of aperture-synthesis radio telescopes, including VPAs, makes it possible to decrease fluctuations of the antenna temperature due to variations in the number of weak sources whose radiation falls into the antenna beam, thereby making it possible to obtain a sensitivity corresponding to the actual antenna area.

Korol'kov and Pariiskii studied the influence of spatial fluctuations in the meta-Galactic background and atmospheric emission on the sensitivity of radio telescopes, and showed that this influence could be appreciably weakened in aperture-synthesis systems during scanning of the antenna beam.

Shivris demonstrated that the surface of an elliptical cone transforms a plane wave into a cylindrical wave, and analytically derived the optimum parameters of the ellipse for which the radial movements of the reflecting elements would be minimised.

Shivris and Pariiskii searched for ways to substantially decrease the systematic errors of the VPA reflector using cylindrical reflecting elements in place of flat ones. In addition, Shivris developed a geodetic method for adjusting the LPRT.

Stotskii and Khodzhamukhamedovii devised an auto-collimation time-pulse method for adjusting the LPRT, which made it possible to arrange the vertically standing reflecting elements in the initial circle with sufficient accuracy.

Gel'freikh proposed a radio astronomy method for adjusting the LPRT using cosmic radio sources, which he then applied together with O. A. Golubchina. This method had the advantage over the auto-collimation approach that it enabled adjustment of the antenna at hour angles close to the elevation of an observed source.

Observations of the radio emission of the Sun at 3 cm with the then unprecedented resolution of about one minute of arc (in one coordinate) began immediately after the geodetic adjustment of the LPRT. Soon, however, the well formed narrow antenna beam began to "fall apart." Analysis showed that the reason for this was in the movable supports of the reflecting elements. A more thorough hydrological study revealed the presence of nearby groundwater sources, and the need to undertake measures to reduce their influence.

Regular observations began in 1957, after the final completion of the LPRT. In 1958, in the group of Kaidanovskii in the Department, Apushkinskii, N. A. Bol'shakov, Butuzov, V. Ya Gol'nev, Pariiskii and V. P. Prozorov developed direct-amplification receivers based on travelling-wave lamps with a 10% bandwidth, designed to operate at 3, 10 and 30 cm with a sensitivity of the order of 0.1–0.2 K and with a time constant $\tau = 1$ s. By this time, at the suggestion of Korol'kov, the first segmented-parabolic reflector was replaced with a feed with a secondary mirror, parabolic cylinder, and a carriage holder for the radiometers and primary feeds, which was installed parallel to the focal line. This enabled the reception of radio emission at several wavelengths (from 3 to 30 cm) simultaneously as the source "images" passed through the focal horns, or to track a point source at one wavelength as it moved due to the motion of the carriage (Fig. 4.4). This substantially improved the sensitivity of the radio telescope.

The use of radiometers with receivers based on travelling-wave tubes in 1957–1958 was progressive. However, the sensitivity of these receivers had already begun

Fig. 4.4 Large Pulkovo Radio Telescope (LPRT)

to limit the capabilities of the LPRT for observations of weak sources and planets by 1960.

Having theoretically analysed the limitations on the sensitivity imposed by sky noise and scattering of radiation by the ground, Korol'kov chose a two-cascade, broadband, parametric amplifier for use in the radiometers, taking into account as well the possibility of cooling these amplifiers in the future. Work on the development of the first parametric amplifier for 10 cm began in 1959 in the GAO Department of Radio Astronomy, under the supervision of Korol'kov. In 1961, a receiver with a parametric amplifier was successfully used in a lunar ranging experiment carried out with the LPRT. A high-sensitivity 4-cm radiometer with a two-cascade, parametric amplifier constructed in 1962 with the participation of Timofeev and Berlin, with subsequent amplification using tunnel diodes, had a sensitivity of 0.03 K for a noise temperature of $T_{\mathrm{noise}} = 150$ K and a bandwidth of 1000 MHz. This radiometer served as a prototype for radiometers operating in various other wavebands.

4.4 Radio Astronomy Investigations

The first radio astronomy observations at Pulkovo began in Spring 1956, on the two 4-m radio telescopes mentioned above. A radio was installed on one of these, which, under the supervision of Khaikin and Kaikanovskii, Shmaonov used to carry out a series of measurements of the radio emission of the sky near the zenith in 1956.

Shmaonov developed a special method for measurements of weak, extended noise emission. The possibility of using a feed horn pointed at the sky as a stable, cold equivalent of the antenna was demonstrated, and was used to develop radiometers for the LPRT.

It also turned out that the sky radiation near the zenith has a brightness temperature of 3.7 ± 3.7 K (maximum error), and the radiation near the pole a brightness temperature of 3.9 ± 4.2 K, which is constant in time. It is now clear that the sky brightness temperature measured by Shmaonov ($\simeq 4$ K) exceeded the temperature of the tropospheric emission in the zenith direction, which modern data show to be approximately 2 K. It follows that Shmaonov appears to have been close to discovering the cosmic microwave background radiation.

Unfortunately, the measurements of Shmaonov, published in 1957 in the journal Pribory i Tekhnika Eksperimenta (Experimental Instruments and Techniques) (No. 1, p. 83), were not known to the Soviet astrophysicists A. G. Doroshkevich and I. D. Novikov, who were the first to point out in 1964 the possibility of experimentally detecting the cosmic microwave background (see Chap. 3). The cosmic microwave background was discovered by A. A. Penzias and R. V. Wilson in 1965, who, in turn, were unaware of the work of Doroshkevich and Novikov, as Penzias noted in his speech on the occasion of their reception of the Nobel Prize for their discovery.

Thus, the work of Soviet scientists included essentially the first, if rather crude and not understood at that time, measurements of the cosmic microwave background (Shmaonov), as well as the first prediction of the possibility of such measurements (Doroshkevich, Novikov).

After its improvement, regular observations of the Sun were begun on the other radio telescope, which was equipped with a polarimeter; these measurements yielded interesting results. It was discovered that the 3.2 cm radio emission coming from certain groups of spots was circularly polarised. Korol'kov, Pariiskii and Soboleva interpreted the polarised radiation emitted by regions above spots as thermal emission by a thin layer in the presence of a longitudinal magnetic field. The circular polarisation was explained by suggesting a difference in the absorption coefficients for the ordinary and extraordinary waves, which correspond to the frequencies $f \pm f_G$, where f_G is the gyrofrequency. It was possible to use the degree of polarisation and the brightness temperature to determine f_G, and thereby the strength of the magnetic field in the corona, which was impossible using optical observations.

It was subsequently elucidated that the proposed mechanism for the generation of polarised emission was appropriate only for regions with thermal radiation. At the same time, as was shown by Molchanov, the magnetobremsstrahlung mechanism also operates above spots, which required some corrections to the magnetic-field strengths estimated earlier.

Molchanov directed his efforts in the Department of Radio Astronomy into regular solar observations over as wide a range of wavelengths as possible. The correctness of his judgement in this was confirmed by the later achievements of the Department in this area.

Two 3-m radio telescopes with polarimeters designed by Kopylov and operating at 4.9 and 2.2 cm were mounted at the Mountain Astronomical Station of GAO, where they were used to carry out regular monitoring of the Sun. Gel'freikh used the results of simultaneous observations of solar radio flares at 4.9, 3.2 and 2.2 cm at GAO and the Mountain Astronomical Station to construct a model for a flare in the form of an expanding coronal condensation with a maximum temperature of 10^7 K. The increase in the degree of polarisation of the flare was explained as the result of a decrease in the optical depth of the condensation as it expanded. This picture is consistent with X-ray measurements carried out later.

The use of polarimeters at different wavelengths during solar-eclipse observations proved to be especially effective. Such studies were begun by Korol'kov and Soboleva in 1956, then repeated many times by various observers, and continue to this day (Gel'freikh, Molchanov, Peterova and others). On many occasions, members of the GAO Department of Radio Astronomy participated in solar-eclipse expeditions, including to China, New Zealand, Mali, Mexico and Cuba, as well as at various sites within the Soviet Union. The Small-Baseline Radio Interferometer was used for some of these observations (Korol'kov, Berlin, Timofeeva, Gel'freikh and others). The structures of local sources were determined, and their polarisation and magnetic fields analysed.

After the operation of the LPRT had been established, the main observations of the Sun, Moon, planets and discrete radio sources were carried out on this instrument.

Centimetre-wavelength observations of the solar radio emission were conducted regularly on the LPRT for a long time. Measurements of the radio-brightness distribution were used to determine the size, temperature and spectral characteristics of active regions on the Sun, together with the dynamics of their development. The radio-brightness distribution of the quiescent Sun was studied as well. In some cases, observations were conducted at two azimuths, making it possible to partially realise two-dimensional high resolution measurements (Ikhsanova, Soboleva, Pariiskii, Gel'freikh).

LPRT observations of the circularly polarised radio emission of local sources on the Sun were of special interest. In such observations, the presence of circularly polarised sidelobes in the LPRT antenna beam represents a source of interference. The first attempt to eliminate this "parasitic polarisation" was undertaken by G. V. Kuznetsova and Soboleva in 1968, but the problem was finally overcome only in the 1970s. It proved possible to make allowance for the parasitic polarisation of the LPRT with high accuracy ($\simeq 1\%$) using two different computational approaches. As was shown by Peterov, the first was based on the fact that the parasitic effect can be excluded through graphical differentiation of the recording of the intensity signal. The second method, developed by A. N. Korzhavin, was based on a superposition of right- and left-circularly polarised antenna beams. Apart from these computational techniques, Esepkina and Temirova developed an instrumental method for removing the parasitic polarisation. After the successful exclusion of parasitic circular polarisation, regular polarisation observations were carried out on the LPRT, facilitating studies of the magnetic fields of spots, including both their strength and spatial structure (Gel'freikh, Peterova, Sh. B. Akhmedov, V. N. Borovik, Korzhavin and others).

In 1959, Pariiskii embarked on an extensive programme of studies of discrete sources on the LPRT at 3, 10 and 30 cm. The coordinates, dimensions, brightness distributions, masses and spectra of a large number of sources were refined, including the nucleus of our Galaxy (work in collaboration with V. G. Malumyan). Thanks to the high resolution of the LPRT, it became possible to identify some discrete sources in other wavebands.

Pariiskii and N. M. Lipovskii investigated variations of the brightness temperature and spectral index of the Milky Way with longitude using the LPRT and several fully steerable radio telescopes with diameters of 3 and 12 m. They derived maps of the Galactic radio emission and estimated the density and mass of ionised gas, together with the number of stars required to excite this mass of gas.

Soboleva used the LPRT to study the linear polarisation of certain discrete sources whose radio emission was non-thermal. In contrast to most polarisation observations, which yielded integrated polarisation measurements for discrete sources, she was able to map the polarisation across the surfaces of some sources.

In 1959, under the direction of Kaidanovskii, Ikhsanova, Shivris and Apushkinskii conducted observations of the Moon with the LPRT at 3 and 2 cm, which confirmed earlier conclusions about the radio-brightness distribution across the lunar disk.

In 1961–1962, Soboleva conducted linear polarisation observations of the lunar radio emission using the LPRT. In observations of the integrated radiation of the Moon using an antenna with a pencil beam, the linear polarisation is eliminated, since the lunar brightness distribution has central symmetry. A knife beam can "cut" a region at the edge of the Moon where the polarised component has a uniform orientation. Soboleva used polarisation observations at 3.2 and 6.5 cm wavelength to determine the dielectric permeability of the lunar surface layer and estimate its roughness.

In 1962, Korol'kov, Pariiskii, Timofeeva and Khaikin observed Venus with the LPRT, using a high-sensitivity radiometer operating at 4 cm wavelength. They were the first to attempt to map the radio brightness across the disk of this planet. The slight limb-darkening that they found was in contradiction with the predictions of the "ionospheric" model, and supported models with a hot surface and cool atmosphere. This result was confirmed several months later by data obtained on the Mariner-2 spacecraft during its flight near Venus (see Essay 1).

LPRT observations of Pariiskii and other researchers in the Department carried out in 1967 (after reconstruction of the radio telescope) at a wavelength of 8 mm, giving a resolution of $15''$, showed that the absorption coefficient for microwave radiation in the atmosphere of Venus was proportional to the square of the density, in agreement with a "steamy" model for the atmosphere. The absence of a dependence of the brightness temperature of Venus on its phase was also established during the course of these observations.

Observations of the radio spectrum of Jupiter in 1964 supported existing concepts about the chemical composition of and the predominant model for the Jovian atmosphere. The parameters of the radiation belts of Jupiter were also estimated.

Preparation for spectral observations in the GAO Department of Radio Astronomy began in 1955. The first task undertaken was to try to detect the deuterium line at 91.6 cm. Under the supervision of Ryzhkov and Egorova, a fixed antenna 15 × 20 m in size with a movable feed was constructed in 1958, for observations at declinations from +84° to +40°. A receiver apparatus capable of integrating a signal over a long time with a sensitivity of 0.5 K was also developed. Before the observations could be done, it was learned that observations using the more sensitive 76-m Jodrell Bank antenna did not detect the deuterium line. Accordingly, it was determined not to be worthwhile to repeat these deuterium observations, but instead to attempt to detect the Galactic line of excited hydrogen at 91.2 cm, corresponding to the transition from the 272nd to the 271st level. The results of these observations showed that this line of excited hydrogen was not detected down to 0.05 K (in terms of its antenna temperature). This implied that the density of exciting ultraviolet radiation was at least a factor of 50 lower than followed from calculations performed by N. S. Kardashev.

In 1959, Rykhkov and Egorova began to develop an apparatus for unexcited hydrogen-line observations using the LPRT, and regular 21-cm observations were begin in 1961, in collaboration with I. V. Gosachinskii and N. V. Bystrova. The goal of the first observations was to learn whether the discrete radio sources near the Galactic centre were associated with its nucleus. Separate spectra of each of the sources in the region of the Galactic centre were obtained, and the distances to the sources were estimated based on their absorption spectra. However, it was not possible to derive an unambiguous relationship between the structure of the absorption lines and the distances to the sources based on these results, due to the strong inhomogeneity of the hydrogen density in the Galactic arms. Further LPRT observations of the neutral-hydrogen line in emission revealed the layered character of the distribution of gas in the disk in the outer regions of the Galaxy. It was discovered that the thickness of one gas layer varies systematically in the transition from the spiral arms to the inter-arm regions, providing evidence for a strong influence of non-gravitational forces on the establishment of the equilibrium of the gas in the disk. LPRT observations also revealed clouds of neutral hydrogen between the spiral arms, which seemed to support the wave theory of the spiral structure. Observations of the 21-cm line in emission and absorption showed that the amount of neutral hydrogen in a dust cloud projected against the Omega Nebula was lower than that in the background gas. Based on this result, it was proposed that there was a transition of gas to the molecular state in this cloud, as was subsequently confirmed by molecular-hydrogen radio observations.

In December 1963, A. F. and Z. V. Dravskikh attempted to detected the excited-hydrogen line corresponding to the transition from the 105th to the 104th level. They detected this line at 5.1 cm in observations of the Orion and Omega Nebulae. The corresponding line at 3.2 cm was detected by R. L. Sorochenko and his co-workers in observations of the Omega Nebula using the 22-m radio telescope of the Lebedev Physical Institute. The results of these studies were presented at the XII General Assembly of the International Astronomical Union in 1964. E. V. Borodzich, A. F. Dravskikh, Z. V. Dravskikh, N. S. Kardashev and R. L. Sorochenko were awarded a diploma for this discovery, dated August 31, 1964.

Over the ten years the LPRT was in use in the GAO Department of Radio Astronomy, more than two hundred observations were carried out. During this time, valuable experience was acquired in the operation and adjustment of variable-profile antennas, and the advantages and disadvantages of the LPRT were studied.

4.5 Search for New Possibilities

In 1955, when the Department was still living on hopes for the LPRT, the development of a much larger VPA-type instrument was begun. Kaidanovskii, Pariiskii, Esepkina, Shivris and Kopylov took part in this work under the supervision of Khaikin. This developmental work served as the initial basis for a number of projects involving large VPAs, including the RATAN-600 radio telescope. Six projects based on very large VPAs were studied, of which three were carried out under the auspices of the Department of Radio Astronomy by very well qualified construction engineers based in both industrial plants and in scientific-research institutes, primarily for applied purposes. A draft design of an international telescope with a reflector area of 10^6–10^7 m^2 and resolution close to that attainable in the optical was developed in the Department.

Possible radio-interferometric instruments were also not neglected. Under the guidance of Khaikin and Kaidanovskii, the Tajik PhD student A. Khanberdiev developed a composite radio interferometer comprised of the LPRT and a two-antenna interferometer with a baseline of 2500 wavelengths (for observations at 9 cm) in 1962–1963, which increased the resolution of the LPRT by a factor of three. The Latvian PhD student A. E. Balklav carried out theoretical studies of the spectrum of spatial frequencies for various aperture-synthesis systems in the GAO Department of Radio Astronomy. In 1963–1964, Kaidanovskii considered the following three types of very-long-baseline radio interferometers: (a) with independent heterodynes in the form of molecular frequency standards; (b) with a common heterdyne placed in a geostationary satellite; and (c) with antennas linked by feed lines in which the phase delays were automatically corrected for.

The urge to probe the limiting possibilities of radio telescopes and radio interferometers in terms of their resolving power and sensitivity was also stimulated by studies of the conditions for the propagation of radio waves in the atmosphere conducted in the group of A. A. Strotskii, in particular studies of the phase structure function and fluctuations of the thermal radio emission of an inhomogeneous atmosphere.

Of all the projects based on large VPAs, only the RATAN-600 radio telescope was realised. This work preceded the reconstruction of the LPRT, aimed at increasing its resolution by a factor of four by reducing the minimum observable wavelength from 3 cm to 8 mm while keeping the other dimensions of the instrument constant.

The reconstruction of the LPRT, which was carried out in 1965–1966, required an analysis of possible ways to increase the resolution and undertaking measures to realise various options. Work conducted by Pariiskii, Shivris, Stotskii, Gel'freikh,

Kopylov, Yu. K. Zverev and other researchers in the Department included (1) the basis for the proposal of replacing the flat reflecting elements of the LPRT with cylindrical elements, and calculating the optimal radius of curvature and dimensions of such elements; (2) developing the technology of forming a precise cylindrical surface of reflecting elements; (3) developing geodetic methods (for studies of the errors and deformations of the reflecting elements, assembly and adjustment of the mechanisms of the reflecting elements); (4) developing radio-technical methods for phase distance measuring techniques for the adjustment of the VPA, and studies of the antenna beam and scattered background for observations at the horizon; (5) development of radio astronomy methods for adjusting the VPA based on observations of the Sun (and subsequently also of the Moon) at hour angles corresponding to the hour angle of a source.

As a result of this reconstruction, the width of the horizontal beam was reduced to 15″ at 8 mm wavelength, which made it possible to map the brightness distribution across the disk of Venus.

Although it confirmed the possibility of increasing the resolution of the LPRT and, in many ways, facilitated the development of the theory of VPAs and of methods for adjusting VPAs, the reconstruction of the LPRT did not lead to substantial new observational results.

It turned out that it was beyond the ability of the Department to construct the required precise secondary mirror for the feed, and the resulting mirror was of good quality but small in size, severely reducing the effective area of the LPRT.

Certain negative side-effects are inevitable in a working group that is large and rapidly growing. Khaikin wrote about these deficiencies in the collection *The Pulkovo Observatory is 125 Years Old*, "Growing pains are inevitable during the rapid growth of a scientific collective—insufficient harmonious coordination of the scientific collective, fracturing of directions of scientific research, the absence of scientific interests that are common to all members of the collective and help to unify them. The battle with these unavoidable difficulties in a rapidly growing scientific collective to some extent facilitated the presence of a single main direction for all the activity in the Department—the development of instruments and observing methods with high resolution at centimetre wavelengths".[3] Indeed, certain work that was begun very early—the development of VPA systems and the improvement of their parameters, studies of their limiting possibilities and the battle to realise these principles—united a large fraction of the group. The experience gained during the reconstruction of the LPRT made it possible to move forward confidently to planning a radio telescope with very large dimensions.

The question of the construction of a VPA radio telescope for millimetre and centimetre wavelengths with a periscope in its southern part and with an area of 2000 m^2 was raised at the initiative of the Sternberg Astronomical Institute of Moscow State University (GAISH) and the Main Astronomical Observatory of the Academy of Sciences of the USSR (GAO).

---

[3] A.A. Mikhailov (ed.), *125 years of the Pulkovo Observatory*, 1966, Leningrad: Nauka (in Russian), p. 57.

On February 5, 1965, the question of constructing this radio telescope was considered and approved at a meeting that included the participation of the President of the Academy of Sciences, Academician M. V. Keldysh, the representative of the Scientific Council on Radio Astronomy, Academician V. A. Kotel'nikov, the Assistants to the Minister of Higher and Middle Special Education, M. A. Prokof'ev and A. F. Metelkin and the Rector of Moscow State University, Academician G. I. Petrovskii.

On March 18, 1965, at a meeting attended by the Director of GAO, V. A. Krat, the Director of GAISH, D. Ya. Martinov, and the representative of the Council on Radio Astronomy, Academician V. A. Kotel'nikov, GAO was confirmed as the leading organisation in connection with the construction of this radio telescope, and the composition and functions of the group to be involved were determined. The radio telescope planned by the USSR Academy of Sciences had a diameter of 600 m, and so obtained the name of RATAN-600 (RAdioTeleskop Akademii Nauk-600). An interdepartamental commission investigated over 20 areas in the Stavropol'skii and Krasnodarskii Territories, and recommended a site at the southern edge of the Zelenchuk Station for the RATAN-600 telescope, 40 km from the Special Astrophysical Observatory (SAO) of the USSR Academy of Sciences, then under construction.

The chosen area was characterised by satisfactory protection from interference and good climatic conditions, and its closeness to SAO provided the possibility of uniting the two scientific collectives around key astrophysical problems. In addition, it became realistic to unify the management and administration of SAO and the RATAN-600.

A technical design for the RATAN-600 telescope was presented by GAO in early 1967.

On June 27, 1967, an expert commission chaired by Corresponding Member of the USSR Academy of Sciences A. A. Pistol'kors approved this project.

In parallel with the development and examination of the technical design of the project, GAO organised cooperation between participants for the planning and manufacture of equipment and the outfitting of the RATAN-600. The supervision of the outfitting of the RATAN-600 was carried out by GAO staff: Pariiskii as the head scientist, Kaidanovskii as the main construction engineer, Korol'kov as the main radio-electronic engineer, as well as assistants to these main positions in charge of various aspects of the work: Esepkina, Shivris, Kopylov, Stotskii, Ryzhkov, Berlin.

A large role in the organisation of the work on outfitting the RATAN-600, including attracting planning and constructional groups to the work, was played by the Commissioner of the Division of General Physics and Astronomy of the USSR Academy of Sciences and senior researcher at GAISH L. M. Gindilis and his colleague from that same Division L. A. Kornev.

Searching for designers of the main equipment of the radio telescope and attracting them to work on the RATAN-600, as well as identifying plants that were able to manufacture the telescope components and carry out the construction and assembly of the telescope itself posed the most difficulties at this stage. Usually, the attraction of new engineering groups began with lectures on radio astronomy and radio telescopes by Pariiskii or Kaidanovskii. These lectures undoubtedly helped involve the heads of various departments and some engineers, which made it easier to obtain agreement about the work to be done from the organisations indicated above. When needed, the management of the Presidium of the USSR Academy of Sciences aided in such matters.

The designing of the main equipment was carried out by the Main Energy Constructional and Mechanical Planning and Design Office of the Ministry of Energy and the Design Office of the Syzranskii Heavy Construction Plant, which also manufactured this equipment.

Simultaneous with the planning and outfitting of the RATAN-600, preparation for its operation and future observations was also begun. Modern radiometers and spectrometers operating at various wavelengths were created by Korol'kov and Ryzhkov.

The group of Stotskii developed methods for radio-technical adjustment of the RATAN-600 and antenna measurements, and also investigated possibilities for its limiting resolution, which were determined by the conditions for the propagation of radio waves in the troposphere. Esepkina and her colleagues refined the theory of VPAs and their electrical characteristics. Zverev participated directly in the geodetic support of the construction and assembly of the RATAN-600 telescope and developed a geodetic method for the precision adjustment of the antenna. The group of Kaidanovskii oversaw the design work, manufacture of equipment and construction and assembly of the telescope components. Astrophysicists under the supervision of Pariiskii continued observations on the LPRT, simultaneously preparing a programme of observations for the RATAN-600 telescope, designed to put its sensitivity and high resolution to the maximum use.

It was not the fate of the great inspirer of the RATAN-600 project S. E. Khaikin to see the fully constructed radio telescope. Already seriously ill, he retired on January 16, 1967, remaining a scientific consultant for the Department. S. E. Khaikin passed away on June 20, 1968.

After the retirement of Khaikin, Pariiskii was chosen as the Head of the GAO Department of Radio Astronomy. This department ceased to exist in October 1969 in connection with its transfer to the Special Astrophysical Observatory, together with its staff and equipment.

The GAO Department of Radio Astronomy was in existence for 15 years. During this time, a capable collective of like-minded radio astronomers attracted by the ideas of Prof. S. E. Khaikin was formed; the main observational instrument of the Department—the LPRT—was developed theoretically and in practise and constructed, and served as an example for the RATAN-600 radio telescope. Hundreds of observations in all areas of radio astronomy were carried out on the LPRT, with resolution and sensitivity that were unprecedented at that time. These observations left appreciable marks in astrophysics. The GAO Department of Radio Astronomy exerted a substantial influence on the work of other Soviet radio observatories, both through its transformation of the PhD students assigned to GAO into specialists in various fields and through its various collaborative scientific work.

Chapter 5
Radio Astronomy at the Special Astrophysical Observatory of the USSR Academy of Sciences

N.L. Kaidanovskii

Abstract The structure of the various radio-astronomy divisions at the Special Astrophysical Observatory (SAO) is described, including the construction and outfitting of the RATAN-600 radio telescope. The results of numerous radio-astronomical observations carried out over ten years on this unique instrument are discussed. The activity of the Leningrad Scientific and Technical Branch of the SAO in the area of radio astronomy is especially noteworthy.

5.1 Formation of Radio Astronomy Departments and Laboratories of SAO and Their Functions

In October 1969, the Radio Astronomy Department of the Main Astronomical Observatory of the USSR Academy of Sciences (GAO) was transferred to the Special Astrophysical Observatory (SAO). At the same time, the Leningrad Branch of the SAO was formed as a basis for the development of promising radio astronomy techniques and methods, as well as the management of observations on the Large Pulkovo Radio Telescope (LPRT). Work intended for this branch included ongoing work on the improvement of the RATAN-600 radio telescope and its receivers and adjustment equipment, development of a system for Very Long Baseline Interferometery (VLBI) observations, preparation of a programme of astrophysical observations and analysis of observations made with the RATAN-600 and LPRT telescopes. Several people under the supervision of G. B. Gel'freikh were left at GAO but transferred to the Solar Physics Department, in order to facilitate the continuation of regular observations of the Sun using the LPRT, and more occasionally the RATAN-600, carried out jointly with GAO.

Yu. N. Pariiskii was appointed the supervisor of radio astronomy work at the SAO and assistant director for research. Carrying out the tasks indicated above required a substantial increase in the number of staff and the organisation of a number of new laboratories.

The organisational structure of the radio astronomy part of the SAO was determined in 1971, and included the following sections.

The **Department of Radio Astronomy Observations** was organised in 1974, immediately after completion of the construction of the first section of the RATAN-600—the northern sector. Pariiskii was selected as the head of this department. The first work of this department was to compose an observational programme aimed at estimating the capabilities of the RATAN-600 in all areas of radio astronomy, from solar to extragalactic.

Further, the RATAN-600 observational programme was determined by an inter-institutional committee. In addition to carrying out the main observational tasks, the Department of Radio Astronomy Observations worked on developing observing methods for the RATAN-600: long tracking of sources, azimuthal aperture synthesis, mapping of the sky, "cleaning" images of extended sources from distortions introduced by the antenna beam and the thermal radio emission of clouds.

A theoretical group was established in the Department of Radio Astronomy Observations and the Radio Spectroscopy Laboratory, whose tasks included the analysis and astrophysical interpretation of observational results and determining bases for future observations. A mathematical group was also established, primarily in connection with the reduction of the observational results and the computation of data to be used when mounting the RATAN-600 reflecting elements.

The **Department of Radio Astronomy Instrumentation** was organised in 1971, with D. V. Korol'kov selected as the head. During the construction and commissioning of the RATAN-600, this department continued the work begun by the group of the main radio-electronic engineer, directly participating in outfitting the radio telescope with its receivers and adjustment equipment. The work of the department included technical adjustments and measurements of the antenna parameters, as well as further refinement of the receivers.

The **Radio Spectroscopy Laboratory** was intended to develop spectroscopic receivers and carry out observations on these receivers. N. F. Ryzhkov was chosen as the head of this laboratory in 1971.

The **Variable Profile Antenna (VPA) Laboratory** was also organised in 1971 under N. L. Kaidanovskii, and carried out the work of the group of the main engineer of the RATAN-600 during the entire period during which the telescope was constructed and assembled. Responsibilities of this laboratory included designing and realising the main and technological constructional elements for the RATAN-600, developing methods for geodetic adjustment, conducting all precision engineering/geodetic work on the radio telescope, further refinement of the reflecting elements and feeds and developing methods and observing regimes for the radio telescope that were compatible with the flexibility of the system.

Starting from October 1, 1973, in connection with the retirement of Kaidanovskii, the responsibilities of the main engineer for the RATAN-600 and the head of the VPA Laboratory were given to O. N. Shivris, with N. A. Esepkina, E. I. Korkin and A. I. Kopylov as his assistants.

The Leningrad Branch of the SAO, which was headed by A. F. Dravskikh starting in 1971, unites workers in all departments and laboratories in the radio astronomy section of the SAO whose work requires that they be either continuously or occasionally located in Leningrad. In addition, VLBI equipment and methods are being developed at the Leningrad Branch in collaboration with the astrometric departments of GAO, and observations are conducted on the LPRT in collaboration with the GAO Solar Physics Department.

In 1977, after the completion of the construction of the RATAN-600, a number of additional laboratories were formed: the **Laboratory for Automation of the RATAN-600 Control System** (headed by G. S. Golubchin), which was primarily concerned with automation of the reflecting-element and feed mounts; the **High-Sensitivity Receiver Laboratory** (headed by A. B. Berlin), working on the development and use of high-sensitivity radiometers, including cooled input amplifiers; and the **Technical Service Group**, which included a mechanical shop for the use, maintenance and repair of equipment.

By 1978, the number of local staff in the radio astronomy sections of the SAO at the Zelenchuk Station had grown to approximately 200.

5.2 Creation of the RATAN-600

The design of the RATAN-600 radio telescope was based on experience with the LPRT and large-scale VPA projects (see Essay 5). However, the RATAN-600 was not a copy of previous projects, and included many new constructive ideas.

In contrast to previous VPA projects, the RATAN-600 has a circular reflecting surface and a flat, periscopic reflector in its southern sector, which facilitates carrying out sky surveys and long signal-accumulation times. In place of a central turning support on which a feed that could move in radius along any azimuth was mounted, the design envisaged moving the feed along one of twelve radial rails laid at intervals of 30°, as a way of simplifying the equipment used and economising on the cost. Transfers between the rails were carried out using a central turning circle (Fig. 5.1).

Calculations made by Pariiskii and Shivris showed that this discretisation of the installation in azimuth does not place any significant limitation on the two-dimensional resolution provided by the telescope in the aperture-synthesis regime.

Since it was planned to manufacture the reflecting elements serially (895 elements 1.94×7.4 m in size for the circular and 124 elements 3.1×8.5 m in size for the flat reflectors), it was necessary to first project, manufacture and test experimental samples of these elements.

As early as 1968, two copies of the reflecting elements for the circular and flat reflecting surfaces were subject to a wide range of tests in the GAO Department of Radio Astronomy. These tests enabled the identification of deficiencies in the drives, position-measuring devices and other nodes. These deficiencies were corrected in the later serial manufacture of the elements.

When planning for the RATAN-600, much attention was paid to the stability of the reflecting elements. The fairly rigid construction of the elements has stood the test of time. The high degree of seismic activity in the region where the radio telescope is located required that each reflecting element be mounted on three spiral supports, which make it possible to rapidly establish the exact location if the foundations have shifted.

Fig. 5.1 RATAN-600 radio telescope. Circular and flat reflectors, radial rails with feeds and the rotating hub at the centre of the circle, as well as the circular paths in front of the flat reflector. The laboratory building and technical workshop are *below and to the left*

At the suggestion of Kopylov, the reflective aluminium sheets making up the elements of the circular reflecting surface were mounted on a girder panel using compressed springs and adjustable screws. To form the sheet, the girders were pressed to a series of profiled template knives located on a special jig. This method enhanced the speed with which the cylindrical surface of the reflecting elements was formed while maintaining sufficient accuracy of the surface.

The elements forming the flat, periscopic reflector (Fig. 5.2) differ from the large circular elements in the presence of a single mechanism for rotating in hour angle and of metallic supports that make it possible to place the panel in an idle horizontal position that does not shadow the southern sector of the circular reflector.

An important role in the realisation of the RATAN-600 project was played by the proposal of Golubchin to use a simple, single-motor drive with an asynchronous motor and a two-stage condensor brake instead of a double-reducer drive with an electromagnetic brake, as was originally planned.

The high accuracy of the lead screws and the reliable means of taking up backlash made it possible to also use the lead screws as measurement devices in the positioning mount system. A measurement-mount device based on double-readout synchro-transmitters connected to the lead screws proposed by Shivris, O. V. Chukanov and Kopylov in 1971 made it possible to mount the reflecting elements with the required accuracy (to within 0.06–0.1 mm) either by hand, using synchro-transmitters outfitted with dials, or automatically, using a computer.

The automated control system for the reflecting elements was based on the use of measurement-mount devices incorporating double-readout synchro-transmitters,

Fig. 5.2 Reflecting elements of the flat reflecting surface on metal supports

which are still used in the present semi-automated control system (with adjustment by hand), and is compatible with this system.

After adjustment, the automated control system for the reflecting elements displayed higher accuracy than in the semi-automated regime, and the time required for the adjustments was substantially shortened.

The RATAN-600 feed consists of a secondary mirror (asymmetrical parabolic cylinder) and a cabin mounted on a trolley supported on three legs. The high-frequency apparatus is located inside the cabin (Fig. 5.3).

The primary feeds are located in a carousel on the focal line of the parabolic cylinder (Fig. 5.4), and can either be mounted on the radio-telescope axis or (to enhance the sensitivity) moved along the focal line, tracking the moving image of the source. The carousel can hold seven to eight primary feeds operating in various wavelength ranges. If the carousel is fixed, the real image of the source will successively pass through the different feeds, making it possible to measure the source flux at the corresponding wavelengths. Another cabin supported on four legs con-

Fig. 5.3 Secondary feed of the RATAN-600 on one of the radial rails

Fig. 5.4 Primary feeds on the focal line of the parabolic cylinder

taining the apparatus for the subsequent amplification and recording of the signals is attached to the cabin described above. Simultaneous observations can be made in three different directions by using three radial feeds, two of which have been used since 1976. A feed that moves along the southern circular track of the radio telescope is constructed in the same way as is described above. The feeds were designed for automated levelling and source tracking.

Designs for new feeds that would enhance the capabilities of the RATAN-600 were finished in 1982. In 1984, a feed for observations of sources near the zenith using the entire circular reflecting surface of the antenna was constructed.

Geodetic work played a large role in the creation of the RATAN-600 telescope, due to the specifics of its construction. Under the supervision of Yu. K. Zverev, SAO geodesists in collaboration with workers of the TsNIIGAiK installation under the supervision of A. G. Belevitin carried out precision geodetic work on the radio telescope and at the plants where the equipment was being manufactured. This work made it possible to avoid appreciable errors, and enabled the RATAN-600 to be realised with parameters corresponding to the design specifications. Geodetic methods for adjusting the antenna were also developed.

Simultaneous with the construction of the RATAN-600, preparation for its operation and the observations to be taken was already begun at GAO. Radiometers and spectrometers operating in various wavelength ranges were designed and constructed under the guidance of Korol'kov and Ryzhkov. A method for radiotechnical adjustment and antenna measurements was developed in the group of A. A. Stotskii, which also studied the limiting attainable resolutions of radio telescopes determined by the conditions for the propagation of radio waves in the troposphere.

Esepkina and her co-workers refined the theory of VPAs and calculated their electrical characteristics.

Astrophysicists under the supervision of Pariiskii continued observations on the LPRT and prepared an observational programme for the RATAN-600, designed to make the maximum use of its special features—high resolution joined with high sensitivity.

Difficulties in financing the expensive instrument provided the main motives behind multiple proposals to lay up the construction of the RATAN-600. With the help of the Division of General Physics and Astronomy (Academician L. A. Artsimovich) and the Council on Radio Astronomy (Academician V. A. Kotel'nikov), the administration of the SAO was able to ensure the continuation of the construction of the RATAN-600, and not allow this work to be shut down. In January 1974, the first section of the RATAN-600—the northern sector (225 reflecting elements) of the circular reflector and one feed—entered its commissioning phase. The entire radio telescope was accepted from the manufacturers in 1977 (Figs. 5.5, 5.6).

The RATAN-600 site now includes a laboratory building with a computer centre, mechanical workshop, cryogenic building with equipment for the production of liquid nitrogen and helium and a small dormitory for the observers and other visitors.

5.3 Beginning of Operation of the Radio Telescope

Immediately after accepting the first section of the radio telescope—the northern sector—from the plant and its preliminary installation and adjustment, various programmes of test observations had already begun by June 12, 1974. It stands to reason that the parameters of the antenna achieved in these first observations were far

Fig. 5.5 Assembly of the reflecting elements of the circular reflecting surface using a special console crane

Fig. 5.6 Northern sector of the RATAN-600 circular reflecting surface

from their projected values. Long, painstaking work to understand the electrical characteristics of the antenna and to identify sources of errors and deficiencies in the reflecting surfaces and their illumination was required. This work was carried out primarily by the Department of Radio Astronomy Instrumentation and the VPA Laboratory.

In order to attain the projected accuracy of the reflecting surfaces, a large amount of work was undertaken concerning the choice of sheet material for the plating and the technology to be used to roll it out, as well as improving the technology for joining the reflecting elements, which enhanced the effective area of the elements by 70% at a wavelength of 8 mm (Shivris, Korkin). In this laboratory, under the supervision of Yu. K. Zverev and with the participation of the Central Scientific Research Institute for Geodesy, Aerial Surveying and Cartography, the errors in the reflecting elements and feeds were analysed, together with the accuracy in the mutual positioning of the elements of the radio telescope and other parameters.

All the engineering and geodetic studies had the goal of working out ways to improve the quality of the reflecting surfaces and introducing corrections to the reflecting-element mounts to take into account kinematic errors in the manufacture and assembly of the mechanisms for the reflecting elements. The VPA methods developed in the geodetic group make it possible to adjust the feed surfaces, determine the radial position of the panels of the circular reflector and position the panels of the flat reflector at transit with a root-mean-square error of the order of 0.1 mm.

The geodetic work on the RATAN-600 demanded the design of a number of specialised measurement devices and gadgets. This work was carried out by the construction engineers of the VPA Laboratory Kopylov and Korkin.

For adjusting the circular reflector using radio-technical means, workers in the Department of Radio Astronomy Instrumentation under the supervision of Stotskii improved the method of autocollimation adjustment that had been developed by Stotskii and N. Khodzhamukhamedov in 1968 in the GAO Department of Radio Astronomy, for adjustment of the LPRT. The distance-phase measurement system used to adjust the RATAN-600 made it possible to place the reflecting elements in the vertical position at the initial circle with an accuracy of 0.06–0.1 mm (in terms of the real mean surface). This system enables estimation of the effective area of the reflecting surface, as well as measurement of the structure of the field near the focal spot, based on the shift of the feed. In this way, Stotskii was able to obtain after adjustment of the reflector a good antenna beam at a wavelength of 8 mm and a satisfactory antenna beam at 4 mm.

In addition to the autocollimation adjustment method, the radio astronomy method for adjusting the reflecting surface developed by Gel'freikh and Golubchina was also applied on the RATAN-600.

During the initial commissioning of the RATAN-600, the horizontal and vertical cross sections of the antenna beams obtained at various hour angles, distant scattering background and noise temperature of the antenna were studied. The characteristics of the illumination of the reflecting surfaces were also measured.

Measurements of the antenna beam near the main lobe and nearest sidelobes, as well as estimation of the effective area of the antenna, were carried out using cosmic point sources. This work was conducted primarily by astrophysicists under the supervision and with the participation of N. S. Soboleva.

An important characteristic of the RATAN-600 is the scattered background, since this background reduces the useful area of the antenna, decreases the achievable dynamic range and increases the noise temperature due to radiation from the surface of the Earth and the atmosphere.

Calculations carried out by B. V. Braude and Esepkina in 1980 showed that the RATAN-600 scattered background has components with directivity determined by both the width of a reflecting element ($\simeq 2$ m) and errors in the manufacturing of the reflecting surface. These errors have a correlation radius of $\simeq 20$ cm, equal to the distance between the screws forming the surface beneath the jig.

Measurements of the scattered background were conducted in 1979–1980 by O. I. Krat under the supervision of Korol'kov. These measurements were done using the radio emission of the Moon, whose size ($\sim 30'$) is relatively small compared to the width of the background radiation ($\simeq 200'$), while at the same time being very large compared to the width of the main lobe of the antenna beam. Thanks to this contrast, the background was increased by $\simeq 30$ dB, enabling measurement of the background radiation to a level of 70 dB.

Through measuring the components of the scattered background, it was possible to identify various types of errors in the surfaces and take measures to eliminate them.

Using all the described methods for adjusting the antenna, and after correctly taking into account the kinematic errors in the adjustment mechanisms, the accuracy of the reflecting surface was increased to 0.3 mm (Soboleva, Zverev, Korol'kov, G. N. Pinchuk, Stotskii). This provided a width of the horizontal antenna beam at the -3 dB level of $1.06\lambda/D$ for all wavelengths down to $\lambda = 1.38$ cm, and a fractional useful area of $\simeq 0.57$—in other words, an effective area of more than 1000 m^2 (at a wavelength of $\simeq 4$ cm). These data are in full correspondence with the projected values.

The excess noise temperature of the RATAN-600 after subtracting the sky noise falling into the main lobe of the beam is determined by re-illuminating the reflector in the vertical direction and the gaps between the reflecting elements. In 1980, the illumination was better optimised, the foundation and ground beneath the reflector screened using a wire grid and these gaps were diminished, leading to an appreciable decrease in the noise temperature of the antenna.

Under these conditions, the RATAN-600 noise temperature for an hour angle of $51°$ at a wavelength of 7.6 cm is 11 K. After subtracting the sky noise falling into the main lobe of the beam due to the cosmic microwave background ($T_{CMB} = 2.7$ K) and the atmosphere ($T_{atm} = 2.8$ K), the excess noise temperature was reduced to 6 K, comparable to the values for horn-parabolic antennas.

With this low noise temperature, it became important to lower the losses and noise temperature of the high-frequency tract. With this goal in mind, in 1980 in the Department of Radio Astronomy Instrumentation, V. Ya. Gol'nev, Korol'kov and P. A. Fridman developed an original design for a radiometer with a noise pilot signal, thereby freeing the high-frequency tract from absorbing modulation devices. In this way, the losses in the waveguide tract were lowered to 0.14 dB, and the corresponding noise temperature lowered to 9.5 K. The total noise temperature at the radiometer input, made up of the temperatures of the antenna and tract, was 21 K.

One of the main directions of work in the High-Frequency Receiver Laboratory was the development of new systems and input tracts with lower noise temperatures. One achievement in this area was the construction in 1980 of a radiometer operating at 7.6 cm. This instrument had a noise temperature of 17 K, of which 12 K arose at the liquid-helium-cooled input parametric amplifier, which had a bandwidth of 500 MHz and an amplification coefficient of 23 dB, while the remaining 5 K arose in the subsequent transistor amplifier.

In 1982, 17 radiometers with good sensitivity operating at various wavelengths, starting from 1.35 cm, were used for regular observations, making it possible to carry out very complex multi-frequency research programmes. The development and testing of radiometers was performed by Berlin, G. M. Timofeev, Gol'nev, A. V. Ipatov and other engineers in the Receiver Laboratory.

In the Spectroscopy Laboratory, T. M. Egorova, A. G. Grachev, A. P. Venger and others under the supervision of Ryzhkov developed various multi-channel filter spectrometers with cooled parametric amplifiers. Apart from these filter spectrometers, an acousto-optical spectrometer developed at the Leningrad Polytechnic Institute under the supervision of Esepkina was also tested on the RATAN-600 antenna in 1980.

The computer-control system for the spectrometers fully automates the observation process. A universal spectral complex for observations of neutral-hydrogen, hydroxyl, formaldehyde and water vapour lines, as well as recombination lines, was completed in 1982.

A method for mapping extended sources was developed in the Department of Radio Astronomy Observations under the supervision of Pariiskii. In the case of objects with very large angular sizes, the images obtained using the knife antenna beams in various azimuths and various source position angles must be synthesised. In addition, a method for rapid mapping of a limited region of the sky or of a fairly compact object was developed. This rapid-mapping method is based on the idea of scanning in azimuth by shifting the primary feed along the focal line of the secondary mirror (parabolic cylinder), and changing the hour angle of the beam axis by shifting the entire feed along rails laid out along the radio-telescope axis.

Kaidanovskii in the VPA Laboratory investigated ways to enable long-duration tracking of sources on the RATAN-600, as well as some types of observational methods that make use of the flexibility of the VPA system.

A method for processing recordings of extended objects in order to free them from distortions introduced by the antenna beam was developed in the Department of Radio Astronomy Observations. When synthesising a two-dimensional image using a known antenna beam, it is possible to conduct a mathematical procedure that "cleans" the "dirty" map, iteratively deriving a final "clean" map that is essentially the image that would be obtained if the antenna beam contained only its main lobe. Implementation of this "clean" process and the subsequent correction of the radio images, as well as taking into account parasitic polarisation, requires knowledge of the antenna beams for the four Stokes components of the radiation for observations at various hour angles and with various feed illuminations. It is also necessary to know the properties of any aberrations that arise, in other words, distortions of the antenna beam that occur due to shifting of the feed.

The calculation of VPA beam patterns, and in particular the RATAN-600 beam, is the subject of many studies by Esepkina, A. V. Temirova and others, beginning with early calculations carried out in 1961 up through their last calculations in 1979. VPA beam patterns were also studied by Gel'freikh and A. N. Korzhavin using optical modelling. This very transparent method makes it possible to derive the beam pattern for specific observing conditions relatively quickly.

The methods and algorithms for computer processing of observations developed by V. S. Minchenko in 1978 make it possible to obtain corrected images of extended objects.

During observations of weak, extended sources using high-sensitivity radiometers, appreciable distortions of the source brightness distribution can arise due to the thermal radio emission of clouds, and also fluctuations in the electrical path lengths for particles that penetrate the inhomogeneous atmosphere and are incident on the antenna. These circumstances stimulated studies of the conditions for the propagation of radio waves through an inhomogeneous atmosphere in the Department of Radio Astronomy Instrumentation, and the development by Stotskii, Dravskikh and A. M. Finkel'shtein of reduction methods that can substantially weaken distortions introduced by the atmosphere.

Fridman investigated the possibility of correcting radio images of extended sources that were distorted by fluctuations in the phase of the spatial harmonics using optimal filtration, based on the criterion of minimising the root-mean-square error. The atmosphere was treated like a filter with random spatial phase-frequency characteristics. He showed that accumulating the phases of the spatial harmonics using aperture-synthesis antennas displayed advantages over building up a radio image from observations obtained on a single-dish antenna.

A technique designed to "clean" an image distorted by the thermal radio emission of clouds developed in 1980 by Kaidanovskii under the supervision of Stotskii was based on the experimentally studied high degree of correlation of the atmospheric fluctuations occurring at different wavelengths, and differences of their spectrum from the spectra of most cosmic radio sources.

5.4 Radio Astronomy Observations on the RATAN-600

After the first test observations—of the source PKS 0521–36° at 4 cm on June 12, 1974—the goal of radio astronomy observations on the RATAN-600 was to estimate the capabilities of the telescope in all areas of radio astronomy, from solar to extragalactic. Thanks to the high resolution and large collecting area of the telescope, these observations sometimes resulted in new discoveries. For example, solar studies conducted under the supervision of Gel'freikh led in 1977 to the detection of very weak active regions—pores, which develop into sunspots. The radio observations of pores sometimes preceded their detection by the optical solar service.

It was shown that the early stages in solar activity are initiated by "radio granulation" (previously observed only during eclipses), which can be observed continuously on the RATAN-600 at many wavelengths. Simultaneous optical and radio observations of the solar granulation can be used to estimate the magnetic fields in these formations. A programme of studies of bright solar active regions was also realised, and a method for reconstructing the three-dimensional magnetic field in these regions developed. The structure of the solar wind was studied at centimetre wavelengths by Soboleva and other astronomers, based on the observed scintillation of quasars.

All these solar observations showed that the RATAN-600 could be used to study a wide range of solar phenomena, from the initial phases of activity to the solar wind. With the sensitivity and resolution of the RATAN-600, the radio emission of not only planets but also their moons became accessible to observation.

Using two of the best radio telescopes in the world at the time (the 22-m Pushchino radio telescope and Effelsberg 100-m telescope), it had been possible to detect radio emission from only two of the Galilean satellites of Jupiter—Callisto and Ganymede. In addition to these two, the Jovian moons Europa and Io were also detected by the RATAN-600 in 1975–1976. These bodies cannot be studied using other radio telescopes because they are located too close to Jupiter, whose radio radiation is a thousand times brighter. The narrow antenna beam and small scattered sidelobe background of the RATAN-600 telescope were required to distinguish the emission of these moons from that of Jupiter itself.

Radio observations of the satellites of Jupiter showed that their emission is thermal, with the corresponding temperature being close to the equilibrium temperature determined by the balance between the incident solar energy and the energy radiated by these bodies into the surrounding interplanetary space.

Among Galactic observations carried out on the RATAN-600 before 1981, the most important are studies of the source Sagittarius A, associated with the centre of the Galaxy. In the opinion of Pariiskii and his colleagues, these observations are not consistent with our picture of the structure of the nucleus of the Galaxy based on radio observations obtained using other major radio telescopes around the world. It was thought that the point source at the Galactic centre was a supermassive star or black hole surrounded by a dense, clumpy cloud of ionised gas, against which supernova remnants are projected. The RATAN-600 observations showed that the source at the centre of the Galaxy is surrounded by a halo of ionised hydrogen, HII, having a surprisingly smooth rather than clumpy distribution. This halo is likely associated with a star cluster in this region.

Pariiskii believes that the power and size of the region of non-thermal emission is not consistent with the idea that the radiation from a supernova remnant gives rise to a "radiation belt" around the nucleus. The distribution of mass in the vicinity of the central star cluster is not consistent with the presence of a massive point source (supermassive star or black hole) at the nucleus. The hypothesis that the point source at the Galactic centre is a pulsar is likewise inconsistent with the RATAN-600 observations, in terms of its spectrum, its luminosity (which is much higher than that of a pulsar) and the absence of pulsations with a period greater than 0.5 s. Observations of the Galactic nucleus with the RATAN-600 telescope help reject a number of accepted but not well based concepts, although they do not provide a complete explanation of the nature of the Galactic nucleus.

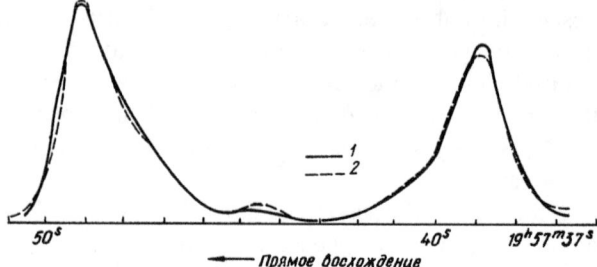

Fig. 5.7 One-dimensional profile of the source Cygnus A: (1) at 6 cm (obtained on the Cambridge radio telescope); (2) at 1.38 cm (obtained on the RATAN-600). The horizontal axis plots right ascension, which increases to the left

The RATAN-600 is characterised by a continuous range of spatial frequencies, from zero to the frequency corresponding to the extent of the antenna in the horizontal direction in wavelengths. Therefore, the radio telescope proved a very suitable instrument for studies of extended objects displaying fine features.

Soboleva used the RATAN-600 to investigate the brightness distributions of 47 radio galaxies over a wide range of wavelengths from 2 to 13 cm in 1977–1980. In contradiction with the established idea that the radio structure of radio galaxies depended strongly on frequency, she showed that the radio structures of nearly all sources studied were approximately independent of frequency. One example is Cygnus A, whose one-dimensional brightness profiles at 6 cm based on data obtained on the Cambridge radio telescope and at 2 cm obtained on the RATAN-600 are very similar (Fig. 5.7). This result demands a re-examination of existing hypotheses about the mechanisms for the formation and maintenance of the radio emission of radio galaxies.

It is possible that the spectrum of a source is formed when the source itself is formed, or that there exists some unified mechanism for the generation of the spectrum in all compact radio galaxies. More than 30% of radio galaxis have a point-like "core" that coincides with the optical galaxy. The spectrum of the core usually differs from the mean spectrum of the radio galaxy. Soboleva traced the evolution of the core spectra of the radio galaxies 3C111, Cygnus A and Centaurus A over 3–5 years. The shape of the spectrum sometimes changed appreciably during the course of a year. For some radio galaxies with "curved" spectra, it was possible to estimate the speed at which components were ejected from the core based on differences in the spectra with time. It was found for Cygnus A, Centaurus A and some other radio galaxies that these speeds were much less than the speed of light.

As an aperture-synthesis system, the RATAN-600 has a high resolution, and therefore a higher limiting sensitivity than other large radio telescopes around the world, such as the 100-m Effelsberg paraboloid and the 300-m Arecibo radio telescope. It's theoretical sensitivity limit at a wavelength of 6 cm is a fraction of a millijansky. After implementing a number of adjustments designed to lower the noise contributed by the antenna and the high-frequency tract, Pariiskii supervised a deep survey of the sky at wavelengths 1.38, 2.08, 3.9, 7.6, 8.2 and 30 cm.

An important instrument that enhanced this deep survey was the radiometer operating at 7.6 cm, which was cooled to liquid-helium temperature. In 1980, using this radiometer, a sensitivity of 0.6 mJy was achieved after an integration time of

Fig. 5.8 Curves showing the dependence of the number of sources N per steradian on the flux density S in Jy (10^{-26} W m^{-2} Hz^{-1}), from measurements with (1) the Effelsberg telescope at 7.6 cm and (2) the RATAN-600 at 6.1 cm. The curve (3) shows the mean curve, and (4) shows the dependence expected for a standard Euclidean Universe

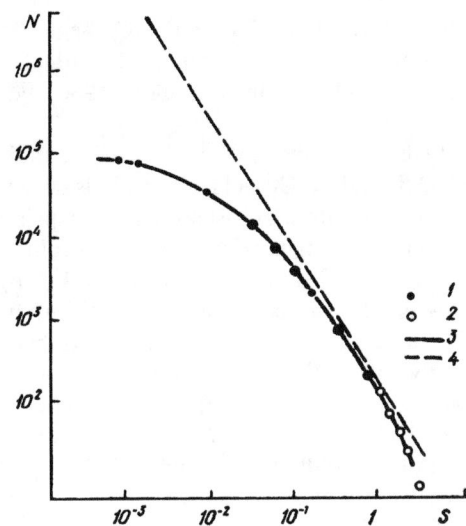

two minutes. The survey was conducted in a strip of sky with a width of ±5′ about the declination of the radio source SS433, covering the full 24 h in right ascension.

Research goals of the survey included determining the behaviour of the number of sources versus flux density ("log N − log S") curve for the weakest detectable sources, studying the cosmic microwave background radiation on various scales and investigating the small-scale structure and magnetic field of the Milky Way. About 3000 sources were detected during the course of the survey, the overwhelming majority of them for the first time.

Preliminary results of the survey showed that the RATAN-600 log N–log S curve agreed well with data obtained on the 100-m Effelsberg radio telescope for flux densities greater than 10^{-2} Jy. At weaker flux-density levels, down to 1 mJy, the growth in the number of sources with decreasing flux density virtually ceases (Fig. 5.8). Pariiskii concluded that the surface density of weak radio sources falls off sharply, and that the number of such sources is 100 times lower than expected for a static Euclidean universe. It also followed that there were no new classes of objects that would dominate over radio galaxies and quasars in this flux-density interval.

A number of attempts were made to use the RATAN-600 to detect fluctuations in the brightness of the 3 K cosmic microwave background radiation. The RATAN-600 antenna beam enables efficient filtration of sky and atmospheric noise within regions of the order of 10′ in size, where theoretical arguments predicted fluctuations in the cosmic microwave background should be largest. However, the data indicated an isotropic background on these scales, in contradiction with the existing theory for the formation of galaxies—a result which has not yet been explained. New measurements at 7.6 cm obtained in 1980–1982 on the RATAN-600 with the high-sensitivity radiometer may help resolve this fundamentally important question.

The Radio Spectroscopy Laboratory of the Special Astrophysical Observatory used the RATAN-600 to carry out a series of observations of lines of HI (1977),

OH and H_2CO (1978) and H_2O (begun in 1981). The entire observation and reduction process is automated. The reduction is carried out on ES-4030 and M-222 computers. Beginning in August 1980, the observations could be processed within a day.

Through a series of studies of the radio source Sagittarius B2 in HI, OH and H_2O lines, I. V. Gosachinskii and his co-workers showed that this source is a young formation with a non-stationary collapsing core. The data obtained were used to construct a dynamical model of the evolution of this interesting source.

In May 1981, Venger, Gosachinskii, Egorova and other workers in the Spectroscopy Laboratory conducted test H_2O line observations of the sources W49 and Orion A. These observations revealed that, during a flare in Orion A, the radial velocity and profile width, and possibly the coordinates as well, varied strongly compared to measurements made in 1979. In December 1981, the same group detected a flare in the H_2O line in W49. N. V. Bystrova and I. A. Rakhimov completed the first part of the "Pulkovo Sky Survey in HI" in 1979; the last three parts were completed by Bystrova and her co-workers in 1980, 1983 and 1084.

In addition to astrophysical observations traditional for radio astronomy, purely astrometric measurements of the precise right ascension of Mercury relative to reference radio sources were carried out in August 1977 with the participation of the Pulkovo astrometrists P. A. Afanas'eva and V. A. Fomin. Due to the narrowness of the antenna beam, Mercury can be observed at virtually all points in its orbit at 2.08 and 3.9 cm, even during conjunctions, when it is quite close to the Sun. The errors for a single observation of Mercury were $\pm 1.4''$, while those for the reference objects were $\pm(0.29-0.73)''$.

Thus, the determination of the right ascension of Mercury conducted on the RATAN-600 at 2.08 cm achieved an accuracy that was close to that of optical observations of this planet. These observations demonstrated the fundamental possibility of increasing the number of observations through multiple experiments, and of distributing them uniformly along the orbits of Mercury and the Earth.

Since the completion of the construction of the RATAN-600 in 1977 and the following period during which its parameters were improved, the scientific programmes and plans for observations have been approved by a programming committee, which distributes the observing time between Soviet and foreign observers; the fraction of observing time granted to the Special Astrophysical Observatory does not exceed 50%.

Together with researchers at the Special Astrophysical Observatory, the RATAN-600 is actively used by groups of astronomers at the Main Astronomical Observatory, the Space Research Institute, the Byurakan Observatory and Moscow State University (GAISH) (see Essays 3 and 6). The radio telescope had carried out more than 10 000 extremely varied observations by 1981.

5.5 The Leningrad Science–Methodology Branch of the SAO

All the departments and laboratories of the radio astronomy part of the Special Astrophysical Observatory had their groups in the Leningrad Branch; in addition, work on the development of methods and instrumentation for a Very Long Baseline Interferometry (VLBI) system were concentrated in the Leningrad Branch. This latter problem was still being examined in the GAO Department of Radio Astronomy, the Lebedev Physical Institute (FIAN), the Radio Physical Institute in Gorkii and the Sternberg Astronomical Institute (GAISH) in 1963–1964. Limitations on increasing the baseline length were considered, and some fundamental possibilities for overcoming these limitations were proposed.

The GAO Department of Radio Astronomy was occupied with the development of large variable-profile antennas at that time, and was not in a position to simultaneously work on a VLBI system. Work on the development of a VLBI system continued at the SAO only in 1969 at the Leningrad Branch, at the initiative of Pariiskii, and in connection with his suggestion to eliminate problems due to phase instability by observing reference sources.

A group supervised by Dravskikh was set up at the Leningrad Branch of the SAO to work on the development of a VLBI system and analyse the possibilities of such a system. In the Spring of 1975, this group completed the development of a narrow-band VLBI instrumentation complex, and test observations were carried out at the wavelength of the hydroxyl line (18 cm) on the baseline between Katsiveli (22-m diameter) and Evpatoria (26-m diameter). In 1978, a VLBI system with a 250 kHz bandwidth ("Mark-I") was completed, and an attempt was made to obtain interference fringes on the Evpatoria–Ussuriisk baseline. In spite of the confirmation that the apparatus used appeared to be operating correctly, neither attempt was successful. The antennas used were made available to the SAO for only a short time, which proved insufficient to learn the reasons for the failure to obtain detectable interference fringes. In subsequent years, work on the development of a wide-band version of the interferometer instrumentation was continued at the Leningrad Branch of the SAO.

V. M. Gorodetskii, Dravskikh and Fridman were able to demonstrate the possibility of synthesising a fairly wide effective frequency bandwidth in the form of a continuous spectrum, and recording this wide bandwidth using narrow-band instrumentation. This is achieved by convolving the spectrum of the signal, together with multiple double-band transformations of the signal frequencies (using both mirror channels). The resulting response has a form that is close to $\sin x/x$, and the threshold sensitivity of the system is appreciably higher than in the system proposed earlier by A. Rogers (USA).

A large series of theoretical works by Dravskikh and Finkel'shtein with various co-authors (1975–1981) was dedicated to analysing ways to realise very high angular resolutions using VLBI methods. A new method for organising VLBI observations and algorithms for processing radio interferometric information (the "arc method") were proposed, which would make it possible to construct an inertial coordinate system, independent of the position of the mean and instantaneous equator of the Earth. The use of this autonomous coordinate system to solve important geodynamical, geophysical and geodetic problems was also investigated. The question of tropospheric constraints in the case of absolute and differential phase and frequency coordinate measurements was investigated in detail using a locally isotropic model

Fig. 5.9 Dmitrii Viktorovich Korol'kov (1925–1984)

for the tropheric turbulence that was developed and refined by Stotskii. Possibilities for using a reference object to calibrate radiometers used in VLBI experiments were also elucidated. The role of special and general relativistic effects in coordinate measurements with high angular resolution was also studied, and the requirements for the structure of an optimal VLBI complex that would be capable of solving a wide range of astrophysical and astrometric problems were formulated. A new algorithm for reconstructing radio images of cosmic radio sources based on the use of "closure phases" was proposed. Dravskikh and Finkel'shtein analysed the relationship between methods for constructing radio images and for determining coordinates using a VLBI array.

V. A. Brumberg (Institute of Theoretical Astronomy of the USSR Academy of Sciences) and Finkel'shtein were the first to fully solve the known general relativistic problem of coordinate conditions, and to identify an invariant procedure for processing observational information in connection with VLBI measurements. They also proposed a number of new tests aimed at verifying the general theory of relativity.

Astrometric research is carried out jointly with the group of radio astrometrists at the Main Astronomical Observatory, headed by V. S. Gubanov. The time service of the GAO took on the duty of providing a means to check clocks used at observing sites during observations.

A substantial push in the development of concepts concerning the capabilities and methods of VLBI was given by the "Poligam" project, which was put forth in accordance with the recommendation of the Radio Astronomy Council in 1978 that an academic, phase-stable VLBI array that could operate 24 hours a day and provide results in near real time was needed in the Soviet Union. The group of Dravskikh actively participated in the development of this project.

When the manuscript for this essay was already at the publisher, the news came of the tragic death on January 10, 1984 of Doctor of Mathematical and Physical Sciences Dmitrii Viktorovich Korol'kov (Fig. 5.9)—the main construction engineer for the radio electronic instrumentation used for the RATAN-600, recipient of a State

Prize of the USSR, participant in World War II. The essays about radio astronomy research at the GAO, and then the SAO, make clear the eminent role played by D. V. Korol'kov in the work carried out in these collectives. Many stages in the development of Soviet radio astronomy were directly related to the activity of D. V. Korol'kov. His death is a heavy loss for Soviet science.

Chapter 6
Radio Astronomy at the Byurakan Astrophysical Observatory, the Institute of Radio Physics and Electronics of the Academy of Sciences of the Armenian SSR and Other Armenian Organisations

V.A. Sanamian

Abstract The establishment and development of radio astronomy in Armenia is described in detail. Information about the radio telescopes of the Byurakan Astrophysical Observatory (BAO) is summarised. The main results of radio-astronomy studies carried out by BAO staff are described, including a number that used large Soviet and foreign radio telescopes, primarily studies of active galaxies.

6.1 Formation and Development of the Department of Radio Astronomy of the Byurakan Astrophysical Observatory, and Its Role in the Development of Radio Physics and Electronics in Armenia

Radio astronomers from the Byurakan Astrophysical Observatory (BAO) of the Academy of Sciences of the Armenian SSR occupy an important position among Soviet radio astronomers.

The Radio Astronomy Laboratory headed by V. A. Sanamian was formed in May 1950, at the initiative of Academician V. A. Ambartsumian. Subsequently, in 1955, it was reorganised into the Department of Radio Astronomy, which is still in existence today. The Department is primarily occupied with research in Galactic and extragalactic radio astronomy and the development of radio astronomy methods.

At the end of 1958, the Very-High-Frequency (VHF) Radio Astronomy Group of the BAO was transformed into the Department of Radio Physics Research Methods, headed by E. G. Mirzabekian. At the initiative of Ambartsumian, the Armenian Institute of Radio Physics and Electronics (IRFE), based on this Department, was established in the city of Ashtarak under Mirzabekian.

In 1968, again at the initiative of Ambartsumian, the VHF Laboratory of IRFE became the Armenian Division of Radio Physics Measurements of the All-Union Research Institute for Physical–Technical and Radio Physical Measurements. In turn, the All-Union Research Institute for Radio Physical Measurements was organised based on this institute, with P. M. Geruni at its head.

As early as 1950, the Mechanical Laboratory was organised in the BAO, which became in 1957 the Optical Mechanical Laboratory, under the supervision of G. S. Minasian. It is this laboratory that was responsible for realising the design and construction of the antennas and antenna-feeder devices for all the radio telescopes of the BAO.

A Department of Radio Physics and Electronics headed by R. A. Kazarian was organised at Erevan State University in 1954 at the initiative of the BAO. Further, in 1975, a Department of Radio Physics was formed at the initiative of Mirzabekian, which prepared specialists in radio physics, including those working in radio astronomy.

Thus, interest in radio astronomy at the BAO and the requirements for its further development was stimulated by the progress of radio electronics and very-high-frequency techniques in Armenia. Each of the organisations indicated above played a large role in the development of science and technology in the Armenian SSR.

6.2 Radio Telescopes of the BAO

Beginning from the first days of the BAO, its director Academician Ambartsumian directed the main efforts of his group into observational astronomy, and radio astronomy was no exception. The BAO Department of Radio Astronomy worked on establishing its own observational base, as well as making use of observational facilities of other observatories. The main focus was constructing radio interferometers, since they provided good resolution and could be comprised of relatively small antennas.

During the first two to three years of the existence of the BAO Department of Radio Astronomy, one radio interferometer after another came into use; these operated at wavelengths of 0.5, 1.5 and 4.2 m, and applied methods for amplitude and phase modulation. By modern standards, they were tiny radio telescopes, but their construction represented a certain accomplishment at the time. They required the development of multi-element, fully-steerable, synchronous-phase antennas, sensitive high-frequency radiometers and acquaintance with new methods for receiving and recording small noise powers.

As we will see below, interesting scientific data were acquired using these radio telescopes. The most effective of them was the two-antenna radio interferometer operating at a wavelength of 4.2 m (Fig. 6.1). It was relocated and rebuilt several times, and was used under both fixed and field conditions. It is still in operation today. The method of phase switching and photographic and resonance methods for integrating weak noise signals were first applied in the USSR on this radio interferometer (see also Essay 1). Data obtained on this radio telescope were used to confirm the presence of secular variations in the radio flux of the supernova remnant Cassiopeia A predicted by I. S. Shklovskii, and to draw the conclusion that the physical mechanisms giving rise to the radio emission of the two brightest radio sources— the Galactic source Cassiopeia A and the extragalactic source Cygnus A—were the

Fig. 6.1 One of the antennas of the first BAO radio interferometer (4.2 m wavelength)

same. This instrument was used to observe partial solar eclipses, to study the propagation of radio waves from cosmic sources through the perturbed ionosphere of the Earth and to acquire interesting data on the absorption and refraction of radio waves in the ionosphere.

In 1967–1968, the BAO radio telescope operating at 1.5 m wavelength was successfully used for systematic observations of the total radio emission of the Sun as part of the programme of the International Geophysical Year.

In 1956, work began on the planning (and, in 1968, on the construction) of a large T-shaped radio interferometer comprised of four antennas whose reflectors had the form of parabolic cylinders, with a total area of 4400 m^2 (Fig. 6.2). This instrument was intended for observations at wavelengths from 0.5 to 4 m, with the change between wavelengths realised by changing the feed system used. The first feed system was designed to operate at 2.5–3.5 m, and was brought into use at the end of 1960. However, strong, regular radio interference appeared in this range in the early 1960's, bringing about the need to move to a new feed system operating at a wavelength of 75 cm. Simultaneous with the change to this new feed system, work was also carried out to increase the area of the radio-interferometer antennas and improve their characteristics. The total collecting area was nearly doubled, to 7800 m^2, and the antennas were equipped with sensitive pre-amplifiers, a system for the calibration of individual sections directly from the pre-amplifier inputs and other devices. All this work was completed at the end of the 1960's. Work on automating the rotation of the radio-interferometer antennas was carried out at the beginning of the 1970's. Note that this instrument has a resolution of 5′ in right ascension and a total effective area of no less than 2500 m^2, which can provide for the mean characteristics of the radiometer a flux sensitivity of 1 Jy for a signal to noise ratio of about five. This radio telescope was successfully used for Very Long Baseline Interferometry (VLBI) observations on the baseline between Byurakan and Simeiz, and, from early 1981, on the baseline between Byurakan and Ootakamoond (India).

The radio telescope is currently undergoing a new stage of reconstruction. One of its antennas with a length of 144 m is being re-outfitted for observations at 327 MHz

Fig. 6.2 Antenna of the large BAO radio interferometer, 216 × 18 m in size (75 cm wavelength)

with circularly polarised receivers, with the goal of using it in a VLBI programme between Byurakan and Ootakamoond, India. A preliminary analysis of the possibilities provided by building a new northern arm to the radio interferometer with a baseline of about 10 km is simultaneously being made. The realisation of such a programme could correct the main deficiency of the radio interferometer—its low resolution in declination. This would appreciably improve the efficiency of the radio telescope, giving rise to new possibilities for using the telescope in various interesting scientific research.

6.3 VLBI Programme

The technique of VLBI opens new possibilities for enhancing the effectiveness of small instruments, which individually are not very sensitive, but can prove to be very useful when they observe jointly with other larger radio telescopes.

Radio astronomers at the Byurakan Astrophysical Observatory were well aware of this circumstance. When at the end of the 1960s the Scientific Council of the USSR Academy of Sciences on Radio Astronomy called for radio astronomy institutions in the USSR to participate in an all-Soviet-Union VLBI programme, radio astronomers at the BAO were among the first to respond to this challenge and to take specific measures to develop this programme at the observatory.

In 1970, jointly with the Gorkii Radio Physical Research Institute (NIRFI), work began on a VLBI system operating at 408 MHz to be used on the Byurakan–Simeiz baseline. The elements of this interferometer were the large BAO radio interferometer and the 22-m radio telescope of the Crimean Astrophysical Observatory (CrAO). A *Razdan-3* computer of the Computational Centre of the Academy of Sciences of the Armenian SSR and a Ch1-48 Soviet frequency standard were used in the system.

As a result of combining the efforts of radio astronomers and engineers at the BAO and NIRFI, as well as staff at the CrAO and the Computational Centre, it was possible to put together a VLBI system for a short time and to conduct observations of a number of quasars. This was essentially the first VLB interferometer in the Soviet Union that was constructed entirely based on Soviet technology. This project was awarded a prize at the Exhibition of National Economic Achievement of the USSR.

In 1972, through the joint efforts of NIRFI, BAO, the Lebedev Physical Institute (FIAN), the CrAO, and IRFE together with other organisations, a new correlator for spectral observations was created, based on the correlator used for the above experiment on the Byurakan–Simeiz baseline. This new correlator was successfully used in VLBI observations of the interstellar water-vapour line at 1.35 cm on the Simeiz–Pushchino baseline.

Unfortunately, however, these collaborative works were not continued after the first successful experiment.

In the last 5–6 years, new steps in developing VLBI have been undertaken at the BAO. At the initiative of the Observatory and jointly with its constant scientific parter, NIRFI, an agreement was reached between the Academies of Sciences of the USSR and India to carry out joint projects on the Byurakan–Ootakamoond baseline (~4000 km). The elements of this interferometer are the Large Ooty Radio Telescope, which has an effective area of about 7000 m^2 at 327 MHz (India) and one antenna of the Byurakan radio interferometer outfitted with a circularly polarised feed and a correlation radiometer based on an ES-1030 computer. This interferometer has already passed the first stage of operational tests on the Byurakan–Ootakamoond baseline.

Participation in VLBI is one of the main and most promising research areas in observational radio astronomy within the BAO Department of Radio Astronomy.

6.4 BAO Participation in Work on the RATAN-600

In addition to observing on the radio telescopes of the Observatory, radio astronomers at the BAO have also successfully used other major Soviet and foreign radio telescopes for observations that are of interest for the general scientific programme of the Observatory. In this regard, a special place is occupied by the largest radio telescope in the Soviet Union, the RATAN-600 of the Special Astrophysical Observatory.

When at the end of the 1970s the Scientific Council on Radio Astronomy called for radio astronomy institutions to participate in the construction of the RATAN-600 telescope in whatever way they could, the BAO was able to make a modest contribution to this project (see Essay 5).

In particular, the BAO and the Armenian IRFE ordered the manufacture of low-noise radiometers with parametric quantum paramagnetic input amplifiers, which were included in the suite of RATAN-600 radiometers and were successfully used for a number of observational programmes, including those initiated by the BAO.

6.5 Development of Radio Astronomy Methods and Instrumentation in the IRFE

Work on high-frequency receivers operating at centimetre wavelengths was carried out in the BAO Department of Radio Physics Research Methods, and then IRFE. A new method for the polarisation modulation of a VHF signal was developed by the then PhD student Mirzabekian under the supervision of S. E. Khaikin and N. L. Kaidanovskii. This method made it possible to distinguish a weakly polarised component of a signal in the presence of a strongly polarised background. The VHF waveguide devices required to realise this method were also constructed: a polarisation modulator, ring feed, balance transformer, broadband phasometer etc. (see Essay 1).

A series of unique VHF radiometers were designed and manufactured at IRFE, and found wide application in studies carried out on space radio telescopes, as well as for space communications and other applied tasks.

The development at the IRFE of quantum paramagnetic amplifiers under the supervision of R. M. Martirosian laid the basis for research in quantum radio physics in Armenia. Efficient quantum paramagnetic amplifiers operating at decimetre, centimetre and millimetre wavelengths were constructed. New methods were proposed for expanding the transmission bandwidth and increasing the inversion bandwidth, based on frequency-modulated pumping. The 21-cm maser developed by Martirosian was the first in the world to be used on the 22-m telescope of the LPI for spectral studies of interstellar neutral hydrogen. The 1.35-cm maser constructed at IRFE is currently included in a suite of radiometers for the RATAN-600 radio telescope, and is being successfully used for spectral observations of interstellar water.

A number of parametric amplifiers in both integrated and band form with fairly good characteristics were developed at IRFE. Radiometers operating at 13 and 30 cm outfitted with parametric amplifiers designed by K. S. Mosoian and R. Kirakosian have been successfully used for many years on the RATAN-600 telescope. F. A. Grigorian was the first in the world to construct a parametric amplifier operating at 4 mm for radio astronomy use.

6.6 Antenna Measurements at VNIIRI

Work on the design and construction of unique antennas for radio telescopes and on antenna measurements was begun in 1968 at the All-Union Research Institute for Radio Physical Measurements (VNIIRI), which had separated from IRFE. A method for engineering calculations of multi-mirror antennas was developed, and applied to radio physical calculations of large radio telescopes in the new system. An original system for the automated control of large antennas that corrects the phase distribution of the field on their apertures was developed. Models of antennas with spherical primary mirrors designed for operation at millimetre and sub-millimetre wavelengths were constructed. A new method was developed for determining the parameters of antennas based on measurements of the field on their apertures, with subsequent machine and optical reconstruction of the field in the far zone. This method is still widely applied both in the Soviet Union and abroad. The construction of a unique two-mirror antenna system with a half-spherical primary mirror with a diameter of 54 m intended for operation at wavelengths from 5 mm to 1 m is ongoing.

6.7 Beginning of Radio Astronomy Observations at the BAO

The first cosmic radio signal at 50 cm wavelength (from the Sun) was detected in May 1951 at the BAO. This was a huge event for the BAO. The still relatively small collective assembled by the modest 3-m radio telescope (adapted from a "Small Wurzburg" radar system), which had been set up right next to the veranda of the laboratory building. Everyone awaited the passage of the Sun through the antenna beam. The spectators were not disappointed—the galvanometer arrow firmly followed the motion of the Sun.

The next step was the reception of signals from the radio sources Cassiopeia A and Cygnus A in Autumn 1951, when one of the antennas of the radio interferometer set up for observations at 4.2 m was ready for operation. A year later, the second antenna of the interferometer was ready, and the first interference signals from cosmic radio sources were received. Regular interferometer observations of roughly the 20 strongest cosmic radio sources began at the beginning of 1953, when a mirror-galvanometer oscillograph was acquired to record the received signals.

Observations of cosmic radio sources were of special interest to Ambartsumian. One antenna with its apparatus cabin was located close to his home at that time, and he very often, usually at night-time, came to the radio telescope to inquire how things were going. One warm autumn night, when Ambartsumian was told that he could see the signal from Cassiopeia A, he immediately came to the radio telescope and requested that the antenna be pointed at and then away from the source several times. He himself turned the antenna, carefully directing it at the source, then ran to the apparatus cabin to see the readout from the instruments. Cassiopeia A is always above the horizon at Byurakan, and this procedure for obtaining the signal went on for several hours. When Ambartsumian was convinced that the signal was real and the readout from the instruments was repeatable again and again, he joyously said, "Now this is happiness!" Note that Ambartsumian was generally very attentive with regard to any observational data.

Another important event was the first observation of a partial solar eclipse, in 1954. The radio astronomers prepared painstakingly for this. At that time, three people worked in the Radio Astronomy Laboratory, which was clearly an insufficient number to carry out observations on three radio telescopes, and it was necessary to mobilise other astronomers and workers for these observations. The efforts were not in vain—good flux-variation curves were obtained on two of the telescopes (operating at 1.5 m and 4.2 m wavelength).

A group of BAO radio astronomers took part in an expedition of the USSR Academy of Sciences to the People's Republic of China, in order to observe the total solar eclipse of 1958. In a short time, a new two-antenna radio interferometer operating at 50 cm was constructed and sent to the site of the expedition. The observations, which were conducted by Mirzabekian, Geruni and G. A. Erzn'kanian, were successful, and yielded interesting data on the radio emission of the Sun, including its polarisation characteristics.

Subsequently, radio observations of cosmic radio sources became commonplace, although, in principle, there are no uninteresting observations.

6.8 Scientific Results

The radio astronomy research programme in the BAO Department of Radio Astronomy as a whole emerged from the general scientific programme of the Observatory. In brief, it consists of studies of non-stationary processes in the Universe, in both our own and in other galaxies. Since the radio emission of cosmic sources can be an important indicator of activity, Galactic and extra-galactic radio astronomy have always been one of the main research directions at the BAO.

In stands to reason that it is not possible in a short survey of the history of radio astronomy to list all the scientific results obtained in the course of three decades of research. We will aim to focus on those that, in our view, are the most important, following a primarily chronological approach.

The first fundamental word on the nature of the radio emission of galaxies was expressed by Ambartsumian. At the Fifth All-Union Conference on Cosmogony in 1955, which was essentially the first radio-astronomy conference in the USSR, A. G. Masevich read a letter written by Ambartsumian, in which he briefly argued against the then popular hypothesis of W. Baade and R. Minkowski that the radio emission of Cygnus A was due to a "head-on" collision of galaxies. Ambartsumian showed that the radio emission of this source, like that of Centaurus A, Perseus A and others, was not a consequence of a collision, but indeed just the opposite—it was the result of processes occurring in the nucleus of the galaxy that were accompanied by the ejection of a huge amount of matter from the nucleus. This conclusion, which formed the basis for a new understanding of radio galaxies, was then laid out in more detail at the Sol'veev Conference (1958), and, in subsequent years, was confirmed by new observational data and became the basis for the theory of the role of activity in the galactic nucleus in the formation and evolution of these galaxies. This theory is currently widely accepted among both Soviet and foreign scientists. Today, it is rare to find anyone who doubts in the presence of active processes in the nuclei of galaxies and quasars, one manifestation of which is their radio emission.

Since we are concerned here with the history of radio astronomy, I think it is appropriate to point out that, at the Fifth All-Union Conference on Cosmogony (1955), Shklovskii, who was then a proponent of the colliding-galaxy theory and presented a talk on the origin of the radio emission in Cygnus A in the collision process, spoke sharply against the comments of Ambartsumian. But a year later, at the following radio-astronomy conference, Shklovskii had clearly moved away from his former position.

One important scientific result of the BAO Department of Radio Astronomy was the experimental confirmation of the secular, monotonic decrease in the intensity of the radio flux from Cassiopeia A, identified with a supernova remnant. Monitoring observations of this source at wavelengths of 1.5, 3.6 and 4.2 m over many years showed, as had been predicted by Shklovskii, that the radio flux of this source over a wide range of wavelengths was decaying almost before our eyes as a result of the expansion of the electron gas in the source and scattering by magnetic fields.

Interesting results were also yielded by observations of the radio emission of the Galactic-centre region (Sagittarius A) obtained at 30 cm and 3.6 m on the Pulkovo and Byurakan telescopes, respectively. These observations showed the presence of fine structure of the radio emission of the Galactic centre, and indicated that the source spectrum was non-thermal, and probably a synchrotron spectrum. The 3.6-m data demonstrated strong absorption of the radio emission at metre wavelengths in regions of ionised hydrogen.

In 1967–1968, members of the BAO Department of Radio Astronomy jointly with radio astronomers of FIAN conducted a series of observations of about 150 radio sources from the 3C catalogue at a frequency of 60 MHz on the East–West arm of the FIAN DKR-1000 telescope using a radiometer specially constructed for this purpose at the BAO (see Essay 1). The low-frequency spectra of these sources were analysed in detail using these data together with data obtained in other studies. These were the first 60-MHz observations for many of these sources, and this frequency is also located near the region where the spectra of many of the sources turn over. For these reasons, these data proved to be very valuable. In particular, analysis of the spectra showed that the spectra of about 60 display low-frequency turn-overs in their spectra, with the curvature of the spectra usually being positive. Low-frequency spectra with negative curvature were more often encountered among double or multiple radio sources.

Interesting studies concerning searches for slow variations in the radio fluxes of a number of quasars and galactic nuclei were carried out at metre wavelengths. Long-term series of measurements of the relative fluxes of the radio sources 3C48, 3C84, 3C273 and others at 408 MHz were carried out over many years at the BAO; the source 3C120 was monitored at 327 MHz on the Indian radio telescope. A number of sources displayed variability at 408 MHz. It was concluded that the observed flux increases were the result of some explosive process. The flux of 3C273 decreased by 25 percent over 10 months, searches for variability in the radio emission of quasars and galactic nuclei are ongoing.

After the role of active galactic nuclei in giving rise to the radio emission of radio galaxies (and then quasars) was understood, a natural question occurred to Ambartsumian: is nuclear activity a general property of all galaxies, including normal, "quiet" galaxies, from the point of view of their radio emission? He presented the scientific staff of the Observatory the challenge of carrying out multi-faceted investigations of normal galaxies, especially those with bluer colours, in order to search for signs of activity in their nuclei.

The very first such searches led B. E. Markarian to the discovery of a class of galaxies with an intense ultraviolet continuum spectrum. Analysing the brightest of these galaxies to investigate the origin of their excess blue radiation, he very early expressed the opinion that this excess radiation was non-thermal in nature. About 1500 such objects were discovered by Markarian and his students over the following 15–20 years, indicating that they form a fairly widespread class among normal galaxies, now known as Markarian galaxies.

Spectroscopic research conducted by E. A. Khachikian and D. B. Vidman revealed a new property of these galaxies. It was found that the vast majority of galaxies with "ultraviolet excesses" in their spectra also have emission lines, and, more interestingly, many display the properties of Seyfert galaxies. This enhanced interest in Makarian galaxies among the radio astronomers of the BAO, as well as among many other radio astronomers both in the Soviet Union and around the world. Since one of the most accessible means to reveal active processes occurring in galactic nuclei is to study their radio emission, radio astronomers, especially those at the BAO, could hardly be uninterested in this class of object.

Over the subsequent 15 years, BAO radio astronomers carried out numerous observations of galaxies with ultraviolet continuum spectra ("ultraviolet continuum galaxies") on various major radio telescopes of the USSR and other countries. For example, G. M. Tovmasian obtained observations of 18 very bright ultraviolet continuum galaxies in 1965 on the 64-m Parkes telescope and the large cross telescope in Australia. The very first set of these observations showed that 80% of these galaxies possess radio emission exceeding 0.1 Jy having a non-thermal spectrum. As we will see below, this large fraction came about due to the selection effect that, as became clear later, this sample included a large number of Seyfert galaxies, whose radio emission is, on average, greater than that of normal galaxies.

In 1970, V. Malumian studied the radio emission of about 30 ultraviolet continuum galaxies by analysing observational data obtained on the Cambridge radio telescope in England. These data showed that the radio emission of these galaxies at 408 MHz did not exceed the sensitivity threshold of this telescope, $\sim$0.5 Jy.

In 1973, Sanamian used the unique Ooty radio telescope in India to observe about 80 selected ultraviolet continuum galaxies. About 10% of these displayed radio emission exceeding 0.6 Jy at 327 MHz. However, due to the large degree of "confusion," this result must be considered uncertain. Some of the firmest results were yielded by lunar occultation data for three ultraviolet continuum galaxies (Mrk384,

Mrk369, Mrk370). In particular, it was found that two of these galaxies (Mrk369 and Mrk384) possess appreciable radio emission at 327 MHz. Based on this result, it was proposed that many galaxies of this class should display radio emission at metre wavelengths exceeding 0.2 Jy if they could be observed with the same resolution and sensitivity as for the lunar-occultation observations with the Ooty telescope. About 100 new ultraviolet continuum galaxies, most of them Seyfert galaxies, were observed on this same telescope by Sanamian and R. A. Kandalian in collaboration with Indian radio astronomers in 1981.

A joint preliminary analysis of the data from these two series of observations confirmed the previous result: of 180 ultraviolet continuum galaxies, 30 displayed radio emission at 327 MHz exceeding 0.8 Jy, with most of them being Seyfert galaxies.

About 500 galaxies from the first five lists compiled by Markarian were observed by Tovmasian and R. A. Sramek on large American radio telescopes in 1973. They obtained a large amount of observational data, enabling a statistical analysis of the question of how the radio emission of these galaxies depends on their morphological type and other characteristics.

When the northern sector of the RATAN-600 radio telescope became fully operational, Sanamian and Kandalian began systematic observations of ultraviolet continuum galaxies over a wide frequency range, from 2.3 to 14.4 GHz. Guided by previous results, they concentrated on those galaxies that also displayed properties of Seyfert galaxies. The number of galaxies observed thus far is 150, with a substantial fraction being Seyfert galaxies. These data appreciably supplemented those obtained by Tovmasian, Sramek and others. Measurements of the radio spectra of a number of galaxies at long wavelengths were continued, and the variability of Mrk348 was confirmed. Some of the most interesting data concern the radio emission of Seyfert galaxies. It was shown that, although the relative number of radio-emitting galaxies is higher for SyII galaxies than for SyI galaxies, the mean radio luminosity of SyI galaxies is appreciably higher (by more than a factor of two to three) than that for Sy2 galaxies. Tovmasian and Sramek had arrived at the opposite conclusion.

Starting in early 1980, radio astronomers from the BAO (Sanamian, Kandalian, G. A. Oganesian) and the radio astronomy station of the Lebedev Physical Institute in Pushchino (V. S. Artyukh and others) began studies of ultraviolet continuum galaxies at 102 MHz using the FIAN Large Scanning Antenna to observe the scintillation of these objects on inhomogeneities in the interplanetary plasma (see Essay 2).

The various results of BAO radio astronomers led to increased interest in ultraviolet continuum galaxies among foreign colleagues, particularly many radio astronomers in the USA.

Malumian obtained systematic observations of about 60 galaxies with high surface brightnesses on the RATAN-600 telescope. He also observed about 140 of these galaxies at 1412 MHz on the Westerbork Synthesis Radio Telescope. These galaxies are also currently being studied at 102 MHz using the Pushchino Large Scanning Antenna (FIAN).

V. G. Panadzhian studied the fine structure of a number of radio sources using observations of their scintillation on inhomogeneities in the interplanetary plasma using the FIAN DKR-1000 radio telescope at 40, 60 and 86 MHz and the BAO radio interferometer at 408 MHz. The characteristics of the scintillation (quasi-period, scintillation index, etc.) were determined as a function of the elongation of the radio source. He demonstrated that the frequency dependence of the maximum scintillation index could be used to estimate the structure of the source. Joint scintillation observations of radio sources at several different frequencies were also used to estimate a number of characteristics of the interplanetary plasma.

6.9 Scientific Connections Between the BAO and Both Other Soviet and Foreign Radio Astronomy Institutions

From the first days of the Department of Radio Astronomy, BAO radio astronomers have been greatly helped by prominent colleagues at the Lebedev Physical Institute, and then the Pulkovo Observatory. BAO radio astronomers carried out their first practical work at the Crimean Station of FIAN in November–December 1950. The first radio astronomer to visit Byurakan for consultation and to offer assistance was V. V. Vitkevich (Spring 1952). With his characteristic enthusiasm, he approved the decision of BAO radio astronomers to concentrate on interferometric methods and recommended that the Observatory work more intensely to develop these methods. Vitkevich remained a great friend and helper of BAO radio astronomers in subsequent years, right to the end of this life.

S. E. Khaikin also greatly aided BAO astronomers. Nearly all the leading BAO radio astronomers (Sanamian, Tovmasian, Mirzabekian, Malumian) did their PhD research under the supervision of Khaikin, with co-supervision by N. L. Kaidanovskii. Khaikin jokingly referred to himself as the "enlightener of Armenian radio astronomers," which was not far from the truth. During a visit to Byurakan in Autumn 1955, Khaikin and Kaidanovskii made a number of important suggestions concerning the subsequent work of the BAO Department of Radio Astronomy.

The positive attitude of these heads of two major radio astronomy centres in the USSR toward the BAO radio astronomers was also maintained by their research groups. The BAO Department of Radio Astronomy has often and effectively collaborated with radio astronomers of FIAN and the Pulkovo Observatory, and there are now collaborations with the Special Astrophysical Observatory as well (Yu. N. Pariiskii, N. F. Rykhkov, A. D. Kuz'min and many others). The collaboration with Leningrad radio astronomers became especially strong after the construction of the RATAN-600 radio telescope began. Currently, nearly all prominent radio astronomers at the BAO carry out systematic observations on the RATAN-600, and have felt the support of the entire staff of the SAO Department of Radio Astronomy, in particular, of its head—Corresponding Member of the USSR Academy of Sciences Yu. N. Pariiskii.

Corresponding Member of the USSR Academy of Sciences V. S. Troitskii first came to Byurakan to organise solar radio observations as part of the programme of the International Geophysical Year in 1976, and a close scientific relationship between the radio astronomers of the BAO and NIRFI in Gorkii has been established since that time. Collaborative work on VLBI has been especially effective.

The scientific relationship between radio astronomers of the BAO and Crimean Astrophysical Observatory (CrAO) is likewise close. In particular, the 22-m radio telescope of the CrAO served as one element in the Byurakan–Simeiz and Simeiz–Pushchino radio interferometers, and CrAO radio astronomers (I. G. Moiseev, V. A. Efanov and others) contributed in many ways to the success of this programme. Moiseev and Tovmasian have also collaborated on studies of radio-source variability at millimetre wavelengths on the 22-m CrAO radio telescope.

The scientific links between BAO radio astronomers and foreign colleagues that were referred to above were not one-sided. Many foreign radio astronomers visited the Byurakan Astrophysical Observatory for extended periods of time and carried out collaborative research with our own researchers (R. A. Sramek and G. Kojoian of the USA; G. Swarup, V. R. Venugopal and M. Joshi of India; M. Kaftan of Iraq and many others). In addition, several dozen foreign radio astronomers have made shorter visits to the Radio Astronomy Department in order to become better acquainted with research at the BAO.

In a number of cases, scientific relations with foreign researchers have developed into ongoing scientific collaborations. A good example is the VLBI programme between the BAO and the Radio Astronomy Centre of the Tata Institute in India, which is carried out as a scientific collaboration in radio astronomy between the USSR Academy of Sciences and the Indian National Academy.

Chapter 7
The Development of Radio Astronomy at the Crimean Astrophysical Observatory of the USSR Academy of Sciences

I.G. Moiseev

Abstract The development of solar radio astronomy on the first radio telescopes constructed at the Crimean Astrophysical Observatory (CrAO) is described. The history of the CrAO 22-m radio telescope is presented, together with the results of important studies carried out on this telescope, primarily at millimetre wavelengths.

Work in radio astronomy at the Crimean Astrophysical Observatory (CrAO) began with studies of the solar radio emission. The radio emission of the Sun is an important, and in some cases the only, source of information about the properties of the solar atmosphere and the physical processes occurring there, which, in turn, influence interplanetary space and determine many geophysical phenomena, such as perturbations of the terrestrial ionosphere and magnetic field.

At the beginning of the 1950s, A. B. Severnii showed through his research of magnetohydrodynamical phenomena on the surface of the Sun that instable plasma formations arise in the solar atmosphere, which should oscillate with frequencies of a fraction of a Hertz or lower. These oscillations could lead to quasi-periodic oscillations in the solar radio emission. He also discovered ejections of hydrogen from the regions of chromospheric flares, which can leave the solar surface at high velocities (~ 800 km/s). Severnii identified these ejections with radio flares displaying slowly drifting frequencies. He also showed that isolated solar radio flares are associated not only with chromospheric flares, but also with the rapid motion of plasma in the solar magnetic fields. As a rule, large chromospheric flares are preceded by isolated radio flares, which he accordingly called "precursor" flares. Radio astronomy observations were not yet being carried out at the Observatory at that time.

To enable more detailed research of active processes on the Sun through studies of their influence on the Earth's ionosphere and magnetic field, the CrAO director Severnii created a special group headed by N. A. Savich within the Observatory, which was located near the Prokhladnoe settlement in the Bakhchisaraiskii region (the Observatory was later given a site in Nauchnii).

Many young specialists joined the Observatory in 1953–1954. These included the radio physicists A. S. Dvoryashin, N. N. Eryushev, I. G. Moiseev, I. N. Odintsova, Yu. N. Neshpor and others.

The Department of the Ionosphere and Radio Astronomy headed by Savich was established at the CrAO in 1954. In that same year, an automated panoramic ionospheric station was planned and constructed (Savich, N. A. Abramenko, V. I. Pylev), as well as other installations. Savich began to investigate the ultraviolet radiation of chromospheric flares based on its interaction with the Earth's ionosphere; Odintsova and Neshpor also studied the behaviour of the ionosphere during flares. Dvoryashin became interested in flows of particles produced by flares and their influence on the Earth's magnetic field, while Eryushev studied the X-ray radiation of chromospheric flares based on its influence on the propagation of long-wavelength noise signals ("atmospherics") from distant lightning discharges.

A group of radio astronomers, initially comprised of only two people—Moiseev and an assistant, P. N. Styozhka—was organised in the Department. The first radio telescope of the Observatory, based on a radar antenna and operating at 1.5 m wavelength, was constructed. The first solar observations began in June 1954. In that same period, much work was done on analysing radio interference and choosing a location for a fixed radio telescope.

A site on the territory of the Observatory near Simeiz on Koshka Mountain was chosen for solar radio observations during the eclipse of June 30, 1954. In the site-selection process, it was elucidated that this was the location of the Crimean station of the Lebedev Physical Institute, where systematic radio observations of the Sun at similar frequencies had been conducted under the supervision of S. E. Khaikin and V. V. Vitkevich, so that information about the occurrence of interference at the site was available. The CrAO radio telescope was constructed as a temporary station, with its receiver located in a transportable cabin, and its mount was azimuthal. The observations of the solar eclipse were successful.

After carrying out these eclipse observations and episodic observations of the Sun at 1.5 m, the need to enhance the stability of the instrumentation became clear. At the end of 1954, Moiseev developed a modulation radio telescope with periodic variation of the shape of the antenna beam, which made it easier to more fully correct for the influence of the Galactic background on the solar emission (Fig. 7.1). Starting in 1955, regular solar observations were conducted on this radio telescope, which, as a rule, were in agreement with optical observations on the new Tower Solar Telescope of the CrAO.

The observations of the sporadic radio emission of the Sun at 1.5 m showed that more detailed studies of non-stationary solar phenomena would require an increase the wavelength range covered by the observations. The planning and construction of a radio telescope intended to operate at 10 cm and a dynamic radio spectrographic designed for wavelengths of 2–3 m were begun.

The work on the radio telescope intended for 10-cm observations began in 1956. By this time the radio astronomy group included, in addition to Moiseev and Styozhka, V. A. Efanov (1932–1983). The strengths of this group were the manufacture of the receiving and recording parts of the instrument. The radio telescope was ready

Fig. 7.1 Co-phased antenna of the first radio telescope of the observatory (1.5 m wavelength)

at the beginning of 1957, and was intended for systematic, continuous (over the entire day) recording of the solar radio emission at 10 cm. With this aim in mind, the antenna was given a parallactic (equatorial) mount, which facilitated tracking of the Sun. A beam-modulation method was developed and applied to decrease the influence of the Earth's atmosphere on the received solar radio emission; this method was based on successively switching at the radiometer input between two feeds located in the focal plane of the 3.5 × 7 m parabolic-section antenna. The commutation of the feeds was brought about using ferrite switches. This radio telescope was late modernised by the radio engineer Yu. F. Yurovskii, and is still used today for regular observations of the Sun.

In parallel with the construction of the radio telescope operating at 10 cm, a metre-wavelength radio spectrograph was also developed and built. A broadband corner antenna uniformly covering wavelengths from 2 to 3 m was constructed for this instrument (Moiseev, A. I. Smirnov; Fig. 7.2). Abramenko built the receiving and recording part of the radio spectrograph, which included electronic adjustment of the wavelength over a time of 0.05 s, making it possible to study the shape of the spectra of rapid solar radio flares.

Fig. 7.2 Broadband corner antenna of the dynamical radio spectrograph

Thus, intense research of the solar radio emission on three instruments in parallel with optical observations had begun at the CrAO by the International Geophysical Year in 1957.

Observations were carried out on relatively simple instruments with modest angular resolution and sensitivity, and more modern instruments were needed for a whole range of radio astronomy research. The fact that construction had begun on the new, precision 22-m reflector of the Radio Astronomy Laboratory of the Lebedev Physical Institute designed by A. E. Salomonovich and P. D. Kalachev (see Essay 1), intended for operation at millimetre wavelengths, had begun in Pushchino in 1959 influenced the process of selecting a design for a larger radio telescope at the CrAO. The first observations on the Pushchino telescope showed its effectiveness for millimetre observations. It was also important that the technical design for the 22-m radio telescope was already available, and experience with its construction and use had already been accumulated. The Presidium of the USSR Academy of Sciences approved the construction of a similar 22-m radio telescope at the CrAO. The organisational work connected with the creation of this instrument was led by Severnii, while Moiseev supervised the construction of the CrAO 22-m telescope.

By the beginning of the construction of the 22-m radio telescope, tests for radio interference had been carried out in various locations in the Crimea. The most suitable place for the radio telescope proved to be a site near the sea to the west of Simeiz (in the Golub Bay). It was sheltered from radio interference on the East by Koshka Mountain, and on the North by a steep ridge of the Crimean Mountains, which also protected the site from wind. The Simeiz region is also characterised by stable clear weather. In addition, territory and residences belonging to the CrAO were located nearby, on Koshka Mountain.

The Observatory's work on the construction of the radio instruments intensified at the beginning of the 1960s. The design and construction of the foundation of the radio telescope and of laboratory and residence buildings was organised by A. G. Pereguda. Work on the assembly of mechanical nodes for the radio telescope was supervised by Yu. G. Monin. In the early stages, much work on the organisation of

7   The Development of Radio Astronomy at the Crimean Astrophysical Observatory

Fig. 7.3 22-m millimetre-wavelength radio telescope of the CrAO

the radio telescope was also carried out by Savich, who was then head of the Department of the Ionosphere and Radio Astronomy. In 1961, this Department became the Department of Radio Astronomy, headed by Moiseev.

The Department of Radio Astronomy designed and built radiometers operating at 4, 8, 13 and 16 mm. These used a multi-frequency modulation method that made it possible to turn on simultaneously at the radiometer input several feeds located near the focus of the antenna. This enabled the localisation of solar radio flares, and accelerated the construction of the corresponding images, while also providing information about the state of the Earth's atmosphere and radio interference during the observations. A small radio telescope (with a 1-m reflecting surface) was installed on a parallactic mound for use in testing the radiometers and gaining experience with millimetre observations (Efanov, Moiseev). Radio polarimeters for observations at 3 cm (L. I. Tsvetkov) and 10 cm (Yurovskii) were also developed and built.

The CrAO 22-m radio telescope was completed in September 1966 (Fig. 7.3). Certain changes had been made to the design of the CrAO telescope, based on the experience gained on the 22-m radio telescope of FIAN in Pushchino. For example, Kalachev and Monin added a central reference axis to the rotating platform, which enhanced the accuracy of the motion of the radio telescope in azimuth. During the assembly of the radio telescope, special attention was paid to the installation of the reflector panels using a knife template. The care put into the installation and verification is demonstrated by the fact that the adjustment of the panel positions took about half a year, and the adjustment of the mirror mounted on the rotating platform about the same amount of time. The panels were positioned to within 0.31 mm (i.e., this is the maximum deviation from the theoretical paraboloid of revolution).

A fundamentally new automated control system run by a computer was developed and realised for the radio telescope (V. N. Brodovskii, V. D. Vvedenskii, N. N. Boronin, Moiseev, I. I. Pogozhev, Yu. N. Semenov, N. M. Yakimenko). A copyright was obtained for this system, which enables pointing of the radio telescope to within $\pm 15''$. A source can be tracked with an accuracy of several arcseconds.

The successful construction of the CrAO 22-m radio telescope was in many ways aided by the constant attention of Severnii and the efforts of P. P. Dobronravin.

A new page in the activity of the Department of Radio Astronomy was opened with the activisation of the CrAO 22-m telescope in 1966. The following main research directions have been followed:

(1) studies of the radio emission of the Sun (mainly non-stationary processes) over a wide range of wavelengths, right down to short millimetre wavelengths;
(2) observations of Galactic and extragalactic radio sources at millimetre wavelengths, including studies of their variability;
(3) investigations of the fine structure of compact sources using Very Long Baseline Interferometry (VLBI), jointly with a number of Soviet and foreign organisations.

In 1963, two laboratory buildings were constructed at the site of the 22-m radio telescope, whose construction had begun, and the Department of Radio Astronomy was moved from Nauchnii to Katsiveli. Soon, the Department had grown to 30 people, and a group had been organised for the administrative and managerial work of the Department. Scientific and technical groups also formed in the Department: the millimetre group (headed by Efanov); short centimetre- and metre-wavelength group (headed by Eryushev); centimetre- and metre-wavelength group, including solar monitoring at 10 cm (headed by Yurovskii); control group for the 22-m radio telescope (headed by I. V. Ivanov, then Styozhka); mechanical workshop (headed by Monin, then M. M. Pozdnyakov); and the optical solar group (headed by P. V. Matveev).

Simultaneous with the construction of the 22-m radio telescope and the manufacture of its radio astronomy instrumentation, studies of the solar radio emission were carried out in the Department of Radio Astronomy using the small radio telescopes referred to above.

Observations of solar flares at 10 cm and 1.5 m and with the radio spectrograph, together with recordings of abruptly initiated magnetic storms at the Earth, were used to investigate the velocities and some other properties of flows of particles ejected from the regions of chromospheric flares, both in the solar atmosphere and in the path from the Sun to the Earth (Moiseev).

Studies of the structure of solar radio flares at 10 cm showed that impulsive flares and the more gently rising and falling flares that follow them are generated in different places in the solar atmosphere. The post-flare decrease in the intensity of the S-component radio emission compared to its pre-flare level, observed in a number of cases, is associated with screening of the ejection sources by material in the flare region (Yurovskii).

The relationship of proton flares to noise storms and powerful solar radio flares were studied, as well as the relationship between solar cosmic rays and chromospheric flares (L. S. Levitskii). The properties of solar noise storms at 1.5 m, and the relationship between the sources of noise storms and dynamical solar active regions in the optical were also investigated (L. I. Yurovskaya).

The small 1-m radio telescope was used to study radio flares at 8 mm. For example, it was shown that, as is true at centimetre wavelengths, three types of flares are observed at this shorter wavelength—impulsive flares, slowly growing and decaying flares and impulsive flares followed by a slow decay. The intensity of impulsive flares at millimetre and short-centimetre wavelengths, on average, grows with distance from the central solar meridian toward the limb (Efanov, Moiseev).

Let us briefly consider some results obtained on the CrAO 22-m radio telescope. 8-mm observations of the radio-brightness distribution over the solar disc were carried out in 1966, with simultaneous mapping of the local magnetic fields on the CrAO Tower Solar Telescope. These showed that these magnetic fields are correlated with inhomogeneities in the solar radio-brightness distribution. Somewhat later, through an analysis of 8 mm observations of the passage of Mercury across the solar disc, it was possible to distinguish fine structure in the radio emission across the solar disc. Two groups of inhomogeneities were detected—one 1.6–2.5″ in size, corresponding to the dimensions of granules, and one 17–40″ in size, close to the dimensions of supergranules.

An increase in the radio brightness at millimetre wavelengths (3.5–13.5 mm) was discovered in polar regions, which appreciably exceeds the radio emission of the unperturbed Sun. The maximum contrast is higher in years of minimum solar activity; that is, the polar regions are more active at millimetre wavelengths at the minimum of the solar activity cycle.

Regular oscillations of the degree of circular polarisation of the solar radio emission at 13.5 mm with a period of 160 min were discovered, which are due to global pulsations of the Sun with this same period.

Eclipse observations at 13.5 mm showed that the sources of the S-component of the solar radio emission are associated with individual spots in a group, with angular sizes 8–30″, effective temperatures reaching 3×10^5 K and degrees of circular polarisation of 15%. A pre-flare lowering of the brightness of the S-component sources is observed at millimetre wavelengths. This can probably be explained as a result of pre-flare heating of optically thin local sources (Severnii, Efanov, V. A. Kotov, Moiseev, N. S. Nesterov).

A multi-mode receiver based on n-InSb crystals developed and built at the Institute of Radio Physics and Electronics of the Ukrainian SSR (A. N. Vystavkin and others) and installed on the 22-m radio telescope was used to obtain observations of the solar radio emission at 0.8–1.2 mm. These showed that there exist appreciable local sources with enhanced radio emission at these wavelengths. The circular polarisation of the S-component sources as they crossed the solar disc was studied at 8 and 13.5 mm. It was demonstrated that S-component sources with circular polarisation with different signs are associated with different parts of a group of sunspots (the leading and trailing regions in the group). The sign of the circular polarisation corresponds to the predominance of the extraordinary wave, and does not change when the sources move across the central meridian (Efanov, Moiseev).

The millimetre spectra of S-component sources from 1 to 17 mm were studied jointly with radio astronomers at the Radio Physical Research Institute in Gorkii. At some wavelengths, it was possible to measure the temperature of the unperturbed Sun using comparisons with the temperature of the Moon (Efanov, A. G. Kislyakov, Moiseev, L. I. Fedoseev).

Regular polarimetric observations of the Sun were carried out at 1.9–3.5 cm. Good agreement was found between variations of the degree of polarisation of the radio emission of local sources and the magnetic-field strength in the corresponding sunspot group. Analysis of the local sources showed that their radio flux densities at 1.9–3.5 cm are most strongly correlated with the visible area of the sunspot group, providing evidence that the radio emission is directed, and that an appreciable role is played by the optical depths of the sources at these wavelengths.

Quasi-periodic fluctuations in the intensities of radio flares at centimetre wavelengths were discovered. Measurements of the solar radio emission at 1.9–3.5 cm revealed fluctuations in the radio brightness with a period of 160 min, which occur synchronously with global 160-min pulsations of the solar surface (A. F. Bachurin, Dvoryashin, Eryushev, Kotov, Severnii, Tsvetkov).

Solar-eclipse observations yielded new data on flare sources and the noise-storm background at 1.37 m. The brightening of the solar limb at 10 cm was also studied in detail (Yurovskii, Yurovskaya).

Studies of discrete cosmic radio sources on the CrAO 22-m radio telescope began at millimetre wavelengths in 1967. The first quasars to be observed, 3C273 and 3C279, displayed strong variability at 8 mm. Outfitting the radio telescope with quantum paramagnetic amplifiers at 8 and 13.5 mm (V. B. Shteinshleiger and others) made possible systematic observations of radio sources with enhanced sensitivity at these wavelengths. A survey of more than 200 sources, most of them extragalactic, has now been carried out. Variations on a time scale of several days were discovered in the quasar 3C273. The reality of these variations was tested in several cases by obtaining observations simultaneously on the CrAO 22-m telescope and either the Parkes 64-m telescope in Australia or the 14-m Metsahovi telescope in Finland. These observations confirmed the reality of these variations. Joint observations on the 22-m radio telescope and the 2.6-m optical telescope of the CrAO showed that rapid variations in the 13.5-mm radio emission of 3C273 were accompanied by a rotation in the plane of polarisation of the optical emission. Rapid variations of the circular polarisation of 3C273 in the optical were also discovered. Variations at 8 and 13.5 mm with characteristic time scales from several months to several years were observed for some quasars. In particular, 3C273 displayed decaying variations with a quasi-period of about a year superposed on slower variations with a characteristic time scale of about 10 years. Studies of the spectra and intensities of the variable components of the radio emission of a number of quasars demonstrated that, on average, the maximum amplitudes of these variations occur somewhere in the wavelength interval from 13.5 mm to 3 cm (Efanov, Moiseev, Nesterov, Severnii, G. M. Tovmasian, N. M. Shakhovskii).

In 1969, the CrAO 22-m radio telescope began to participate in international VLBI observations aimed at investigating the fine structure of compact radio sources. A record resolution for a ground-based radio interferometer (0.0004″) was achieved in 13.5 mm observations on the Crimea–Haystack (Massachusetts) baseline. VLBI observations were also used to study the fine structure of Galactic sources in the water-vapour line at 13.5 mm. In 1980, the radio interferometer baseline Crimea (Crao 22-m)–Pushchino (FIAN 22-m) was used to investigate the fine structure of various compact radio sources at 13.5 mm (Vitkevich, Efanov, V. I. Kostenko, L. R. Kogan, L. I. Matveenko, Moiseev, R. L. Sorochenko, I. S. Shklovskii).

In the CrAO Department of Radio Astronomy, much attention was paid to automating the observation and data-reduction process. For most observations, the control of the 22-m radio telescope and the reduction of the data obtained were carried out in real time using an M-6000 computer, substantially enhancing the efficiency of these studies.

The Department of Radio Astronomy is involved in ongoing collaborative research on a number of current problems in astrophysics, with radio astronomy centres both in the USSR and abroad, in Australia, Great Britain, Cuba, the USA, the Federal Republic of Germany and Finland.

Chapter 8
The Development of Radio Astronomy Research at the Institute of Radio Physics and Electronics of the Academy of Sciences of the Ukrainian SSR

S.Y. Braude and A.V. Megn

Abstract The history of the development of radio astronomy in the Ukraine is described. The construction of unique radio telescopes and radio interferometers operating at decameter wavelengths is of special interest. The results of important studies of the Sun, planets and Galactic sources at these wavelengths are presented. The importance of detecting atomic line radio emission is underscored.

8.1 History of the Development of Radio Astronomy in the Academy of Sciences of the Ukrainian SSR

Radio astronomy research at the Institute of Radio Physics and Electronics of the Academy of Sciences of the Ukrainian SSR (IRFE) began with studies of the propagation of radio waves with various wavelengths (from Very High Frequency to medium-wave) above the interface between two media and in a plasma. This work was first carried out in the Department of Radio-Wave Propagation, which was created in 1945 in the Physical Technical Institute of the Academy of Sciences of the Ukrainian SSR under the scientific supervision of S. Ya Braude. Beginning in 1955, when IRFE was organised, based on the Physical Technical Institute, this work was continued in three departments of the new institute. The most noteworthy of these studies conducted at that time (Braude, I. E. Ostrovskii, Ya. L. Shamfarov, I. S. Turgenev, A. V. Megn, A. I. Igrakov, V. F. Shul'ga, O. N. Lebedeva and others) include studies of the propagation of short-, intermediate- and medium-wavelength radio waves above the surface of the sea and the scattering of such waves by irregularities in the sea surface. This led to the emergence of a new area of research, called radio-oceanography, which is still being developed today.

Radio-oceonographic studies required directive antennas to radiate signals and then receive the scattered signals from particular areas of the sea surface, with the possibility of rapidly changing the direction of the antenna beam in space. In contrast to centimetre, decimetre and metre wavelengths, which were used for radar at that time, the need to develop electrical rather than mechanical methods for directing the antenna beam of a radiating system arose for short- and more long-wavelength radio waves. At that time, only one highly-directive short-wavelength antenna was known: the "Musa" antenna, with electrical pointing of the beam in hour angle, which was developed in the 1930s for short-wave communications between the USA and England. Research carried out by scientists in the Department of Radio-Wave Propagation (Braude, Megn, Ostrovskii, V. S. Panchenko, Yu. V. Poplavko, Shul'ga) concerned antennas designed for the directed radiation of signals and the reception of such signals scattered by the sea surface operating at short and intermediate wavelengths. Some of these antennas already incorporated electrical pointing of the beam in azimuth. The principles of antenna construction and methods for battling with various types of decametre-wavelength interference that were worked out during this research were subsequently used in the development and construction of a series of decametre-wavelength radio telescopes of the Radio Astronomy Division of IRFE.

Simultaneous with studies of the propagation of radio waves above an interface such as the sea surface, the conditions for the transmission of such waves through a plasma such as the solar corona were also considered. This question is relevant in relation to possible radar sounding of the solar corona. Work dedicated to the idea of radar studies of the Sun was carried out in 1957 (F. G. Bass, Braude). Computations suggested that such studies could be successfully carried out if frequencies from 10 to 40 MHz (decametre wavelengths) were used. A computational effective area for the receiving and transmitting antennas of about 10^5 m^2 was required for a mean transmission power of 10^5–10^6 Watt. This spurred on the desire to carry out work in radio astronomy at decametre wavelengths; the question of designing highly directive antenna systems operating at these wavelengths capable of tracking various cosmic objects was simultaneously addressed. Given the difficulties in building such a powerful transmitter, it was decided in 1957 to first construct only the receiving part of the system, which could also be used for a variety of research in observational decametre-wavelength radio astronomy. A new scientific department was organised at IRFE in 1958—the Department of Cosmic Radio Astronomy, headed by Braude. In that same year, IRFE was given a plot of land 140 hectares by 80 km in size to the southwest of Khar'kov near the Grakovo station, to be used for experimental research.

The main task of the newly created department was the final selection of the main research directions to be pursued, the establishment of a radio astronomy observatory and carrying out observations for various programmes. As is noted above, one possible research direction was connected with decametre-wavelength radio astronomy. A small number of such studies were carried out in Australia and England at the end of the 1960s. In a number of cases, the results of different studies contradicted each other, even those for the most powerful sources (Cassiopeia A, Cygnus A, Sagittarius A, Virgo A). At the same time, numerous data on discrete radio sources, the Galactic radio background, neutral hydrogen, and the Sun, Moon and planets had been obtained at metre and decimetre wavelengths. Together with this direction, the possibility of conducting observations at higher frequencies was also investigated. It was obvious that it would be fruitful to pursue radio astronomy research at metre and shorter wavelengths only if IRFE could acquire an instrument whose capabilities were no lower than those possessed by radio telescopes belonging to other observatories.

By the end of the 1950s, the Large Pulkovo Radio Telescope (see Essay 4) had been built, as well as the fully steerable 22-m parabolic radio telescope of the Lebedev Physical Institute (see Essay 1). These had begun to make observations at decimetre, centimetre and millimetre wavelengths. Several fairly large radio telescopes had also been constructed outside the Soviet Union, including the 76-m radio telescope of the Jodrell Bank Observatory in England, intended for observations at metre and decimetre wavelengths. Therefore, it was decided in IRFE to develop the design for a fully steerable parabolic antenna with a diameter of 200 m. A draft design for such a radio telescope with an equatorial mount intended to operate at metre, decimetre and centimetre wavelengths was produced at Khar'kov by the constructional engineer S. I. Kuz'min. The approximate cost of developing and realising this instrument was estimated to be more than 150 million rubles.

Simultaneous with planning for this instrument, the possibility of creating a radio telescope with a large effective area suitable for observations at decametre wavelengths (a decametre radio telescope) was investigated. At first, when it was proposed to use the receiver antenna system only for radar studies of the Sun, an antenna in the form of a multi-dipole horizontal square grid 300×300 m in size was considered, which was to be uniformly filled with horizontal turnstile half-wave dipoles. Since radar measurements of the Sun must be carried out at multiple wavelengths, the antenna system was to be broadband, or at least multi-wavelength, with the ability to tune to a number of decametre wavelengths. It was necessary to provide a means for the antenna beam to track the Sun, which, in the case of such a large instrument, could be done only using electronic pointing of the antenna beam. However, when it was decided to also use this radio telescope for a number of other radio astronomy programmes, the chosen square configuration of the antenna proved not to be optimal, since it provided a resolution at 20 MHz of only $3 \times 3°$. An optimal configuration should provide agreement between resolution and sensitivity, or realise the maximum resolution for a given sensitivity. The requirements can be satisfied by a combination of linear antennas having a cross-, T- or Π-shaped form. Based on the reliability of the antenna system and its stability against interference during measurements for various types of programmes, and also the desire to appreciably increase the volume of information that is simultaneously obtained during the observations, an optimal decametre radio telescope should operate over a range of wavelengths; the chosen range of frequencies was from 10 to 25 MHz.

A comparison of the two preliminary telescope designs and a final selection of the research directions to be pursued in radio astronomy was carried out in 1961–1962. The following considerations were the deciding factors. The reflector operating at decimetre and centimetre wavelengths could be constructed only with the involvement of industry, and required the allocation of considerable resources.[1] The participation of the IRFE Department of Radio Astronomy in this work could prove to

---

[1] As the development of radio astronomy has shown, this path was not the only possible one. High resolution can be realised using synthesised apertures, very long baseline interferometers, diffraction techniques, such as the occultation of radio sources by the Moon and so forth; however, at the

be quite limited, since the design, manufacture and assembly of a large parabolic mirror, as well as the mechanical elements and drive of the telescope would have to be done by specialised constructional and machine-building enterprises. The radio-technical work in outfitting such an instrument, in which the Department of Radio Astronomy could actively participate, would be important, but only a small part of the overall project. In addition—and this was the deciding factor—it was unlikely to be possible to obtain the necessary large-scale allocations of finances and resources.

A completely different situation came about in decametre-wavelength radio astronomy. The setting up of the antenna grid itself, consisting of a large number of identical half-wave dipoles that were individually manufactured and installed in an antenna field with comparatively low accuracy, of the order of tens of centimetres (about $\lambda_{min}/10$), could be carried out by a specialised constructional–assembly organisation. However, the main efforts in the creation of such a telescope are associated with the necessary radio-technical development, and the design and manufacture of all the radio systems used—broadband symmetrising and matching devices, high-frequency communications devices, a system for electronic control of the beam, antenna amplifiers, various control devices and so forth. This work could be carried out by staff in the IRFE Department of Radio Astronomy and the experimental facilities of the institute. Estimates of the cost of the proposed decametre radio telescope indicated that it would require about 3 million rubles.

Simultaneous with studies of an effective decametre-wavelength instrument, the potential fruitfulness and possible scientific research directions of this wavelength range were also considered. Decametre wavelengths are the longest for which radio astronomy observations can be done from the surface of the Earth, and are of considerable interest in connection with various mechanisms for the radiation, absorption and scattering of radio waves in space. One such mechanism is the absorption of radio waves in the radiating medium itself, made up of relativistic electrons moving in magnetic fields with velocities close to the velocity of light (so called self-absorption). Analogous effects arise due to distortions of the energy spectra of relativistic electrons at relatively modest energies, the group interactions of electron beams with cosmic plasma observed both in our Galaxy and in the Metagalaxy, the absorption of radiation in clouds of ionised hydrogen and a number of other processes. All these processes can be studied by measuring the spectra of cosmic radio sources at decametre wavelengths.

Measurements of decametre radiation together with data obtained at shorter wavelengths can be used to determine the strength and direction of cosmic magnetic fields, and the distributions of the density and energy of the electrons and other particles. A number of very important physical parameters of normal galaxies, radio galaxies, quasars, supernova remnants, the solar corona etc. can also be derived with such measurements. Almost no such investigations were carried out at decametre wavelengths in the 1950s and the beginning of the 1960s due to a whole range of factors, which proved not to be important at higher frequencies. First and

---

beginning of the 1960s, these methods were not yet widely used, and so were not considered in detail.

foremost, the low resolution due to the comparatively small size of existing instruments made it difficult to distinguish signals from individual cosmic radio sources, leading to so called confusion, which limited information that could be derived from observations.

Measurements at decametre wavelengths are also complicated by the influence of the Earth's ionosphere, which gives rise to absorption, as well as amplitude, phase and polarisation fluctuations of the received radiation. Ionospheric effects essentially make observations at long decametre wavelengths from the surface of the Earth technically impossible. Considerable difficulties are also associated with various types of interference in this wavelength range, especially with signals from a large number of short-wave radio stations, which often exceed the intensity of the radiation received from a cosmic source by many orders of magnitude (by a factor of a million or more). In radio astronomy observations, we must also deal with the fact that the temperature of the diffuse radio emission of our Galaxy (the Galactic background) reaches several hundreds of thousands of Kelvin, while this background does not exceed 10 K at centimetre wavelengths. All these factors indicate that most radio astronomy measurements at decametre wavelengths can be effectively carried out only using a radio telescope with fairly high resolution and sensitivity, making it possible to weaken the influence of some of these hindrances.

As a result of such analyses, it was decided to develop a series of radiometer studies at decametre wavelengths at IRFE. Since there was no highly directive, broadband, decametre radio telescope that was electronically controlled in both coordinates and had an optimal configuration existing at that time, there was the danger that it would prove unrealistic to realise such a project in full form through the efforts of specialised project-oriented organisations. Through detailed discussions of this question at the Scientific Council on Radio Astronomy of the USSR Academy of Sciences, at the suggestion of the Chairman of the Council Academician V. A. Kotel'nikov, it was decided to approve a large T-shaped decametre radio telescope, subsequently named the UTR-2. The development and technical design of this project was assigned to the Institute of Radio Physics and Electronics of the Ukrainian SSR Academy of Sciences.

8.2 Decametre Radio Telescopes of the IRFE Radio Astronomy Division

The Department of Radio Astronomy Antennas and Instruments was established at IRFE to carry out scientific and technical work associated with the construction of a large decametre radio telescope. The head of this department, A. V. Megn, supervised the development and construction of all decametre radio telescopes in the IRFE Radio Astronomy Division. No previous experience with building the unique antennas needed for such a radio telescope was available either in the Soviet Union or abroad. For this, it proved necessary to develop and construct several comparatively small test instruments operating from 10 to 40 MHz simultaneously with the

Fig. 8.1 Antenna arrays of the second version of the ID-1 interferometer

planning of the UTR-2 telescope, to address a whole series of scientific and technical questions. First and foremost, such questions concerned the construction of broadband phasing systems for the electronic control of the antenna beam, methods for calibrating all the main characteristics of such radio telescopes, the provision of the required high stability of the receiver against interference while simultaneously compensating the high losses of the antennas, accurately taking into account the influence of the Earth's surface on the characteristics of the instrument, the development of the principles behind high-frequency control of the antenna, the creation of the necessary baseline elements etc.

The first of the small radio telescopes (the ID-1—an interferometer operating at decametre wavelengths) was constructed in two forms: antenna rails extending to braces (1960), and grids of freely standing radiating elements (1961). The second (main) version of the instrument, which was used for measurements at 12–20 MHz, was comprised of two identical, 24-element, horizontal antenna grids receiving linearly polarised signals separated along a line East–West by 332 m (Fig. 8.1). Each of the antennas had an area of 24×48 m and four rows of six dipoles oriented East–West. All the dipoles were placed at a height of 4.5 m above the ground and were oriented along the interferometer baseline; the distance between the dipole centres along and orthogonal to the rows was 8 m.

The radiators were horizontal, symmetrical, broadband shunt dipoles. Each dipole consisted of two wire cylinders 1 m in diameter and 3.75 m in length, which were attached to a vertical metal mast with metallic isolators (shunts). The radiator signals were summed using a two-stage circuit, and were phased using a one-stage, parallel circuit. The radio telescope was intended for interferometric measurements on a single baseline, to be used as a meridian instrument (with electronic control of the antenna beam in one coordinate—declination). This radio telescope already made use of antennas consisting of broadband dipoles, with a well matched, screened, parallel circuit for feeding the radiators and very simple broadband sys-

Fig. 8.2 Western ID-2 antenna

tems for matching the input resistances of the radiators and the wave impedance of the antenna, and with electronic control of the beam via hand switching of the time delays in the rows of feeds. However, due to insufficiently good matching between the radiators and the phasing system, and also due to the lack of good isolation for all the high-frequency communication elements, the ID-1 could only approximately be considered a broadband instrument, since its characteristics were very wavelength dependent within the operational wavelength range. The ID-1 radio telescope was immediately followed by the construction of the more modern ID-2 (Fig. 8.2).

The ID-2 radio telescope, built in 1962, was able to carry out measurements at 20–40 MHz. It consisted of two rectangular antenna grids 16.5 × 176 m in size, separated by 470 m along a line East–West. These grids, which each contained 128 broadband radiators, were orthogonal to each other: the western antenna consisted of four rows oriented North–South with 32 parallel dipoles in each row, while the eastern antenna had four rows oriented East–West with 32 co-linear dipoles in each row. The orthogonal mutual orientation of the antennas, each of which was highly directive only in one plane, provided high directivity for the radio interferometer in two planes, corresponding to a pencil beam. The radiators in both antennas were symmetrical, horizontal shunt dipoles 1 m in diameter and 4.75 m long fixed at a height of 3 m above the ground. The distance between neighbouring radiators along and orthogonal to the rows was 5.5 m.

Like the ID-1, the ID-2 radio telescope was initially a meridian instrument, with electronic control of the beam only in declination.[2] In both of the ID-2 antennas, the radiators were fed using multi-stage, matched, unisolated circuits. In the western

---

[2] The ID-2 was subsequently modernised to enable simultaneous measurements on two baselines (470 m and 88 m) by dividing the eastern antenna into two identical sections. In this case, the eastern antenna began to also be phased in its second coordinate—hour angle.

antenna, there was first a cophased summation of the signals of the four colinear dipoles (along the short rows from West to East) in 32 identical adders. The electronic pointing of the beam at one of 256 possible positions in a ±90° sector in zenith angle was carried out using a three-stage temporal phasing system consisting of 21 phase transformers, through which the short rows of dipoles were fed. In the eastern antenna, there was a cophased summation of the dipole signals in the rows (from West to East) using a three-stage parallel circuit, with every four added in the first two rows and every two added in the last two. The signals of the rows were phased using a single phase transformer, providing 32 positions of the beam in a ±90° sector in zenith angle. The phasing apparatus for these antennas consisted of discrete temporal delay lines constructed according to the duality principle, that could be switched using high-frequency relays that were controlled remotely. An economical asynchronous means was used in the control apparatus for the beam of the western antenna, when the number of beam positions (multiplicative factors) for various stages of the phasing was chosen to be different—from the minimum for the least directed multipliers for the first stage of the phasing system, to the maximum for the beam of the entire antenna. To optimise the characteristics of the radio telescope, the ID-2 antennas, like the ID-1 antennas, used simple broadband matching of the input resistances of the dipoles and the wave impedance of the system for phasing and feeding the receivers. To reduce the influence of losses on the receiver sensitivity, antenna amplifiers made according to the standard broadband scheme were implemented at the antenna outputs. Although there is no doubt that this radio telescope was more efficient than the ID-1, it had insufficient resolution and sensitivity, and it was not able to provide the accuracy in the calibration of the main characteristics of the antennas required for precision absolute measurements, due to insufficient isolation in the chains for feeding and phasing the dipoles, as well as non-optimal matching with the phasing system.

The third-generation UTR-1 (Ukrainian T-shaped Radio telescope, first model), developed and built in 1963–1964, was appreciably more modern. Its main characteristics were studied in 1965, and the UTR-1 was introduced into regular use beginning in 1966. This instrument, which operated at 10–25 MHz, had a T-shaped configuration that was close to optimal from the point of view of matching its resolution and sensitivity. Its beam was highly directive in two planes, and was formed by multiplying the antenna beams of the two mutually perpendicular antennas, each of which was highly directive in only one plane. The first antenna of the URT-1 was oriented North–South, had a length of 600 m and was comprised of 80 dipoles forming a single row. The second antenna was oriented East–West, had a length of 576 m and was comprised of 128 dipoles forming two rows. Both of the UTR-1 antennas was designed to receive linearly polarised radiation, and was made of individual broadband, horizontal, symmetrical shunt dipoles with a diameter of 1.8 m and a length of 8 m. All the dipoles were fixed at a height of 3.5 m above the ground and were oriented East–West. The dipoles were all separated by 9 m East–West and 7.5 m North–South, including the nearest dipoles in the East–West and North–South antennas (Fig. 8.3).

The radio telescope beam was controlled electronically in both coordinates at a distance: in zenith angle within ±80° and in hour angle within approximately

Fig. 8.3 North–South antenna of the UTR-1 radio telescope, comprised of 80 oscillators

±1.5 hours. Broadband, discrete (on temporal delay lines) phasing systems were used for this purpose: two to control the antenna beams and a third for their mutual phasing. A six-stage asynchronous phasing system providing 512 beam positions in zenith angle was used in the North–South antenna. In the East–West antenna, the beam control in zenith angle was carried out using a one-stage circuit that provided 16 beam positions, and the control in hour angle using a five-stage cophased circuit with seven beam positions.

Broadband, hybrid, directive devices were used in all the UTR-1 phase rotators and signal adders, which provided good isolation between all the high-frequency channels. Thanks to this, all of the radiator tracts displayed virtually no reaction to the other high-frequency chains of the antennas, which helped ensure good stability of their characteristics and high reliability in their operation. To optimise the characteristics of the radio telescope (decrease their frequency dependence and realise the maximum sensitivity), broadband matching of the input resistances of all the radiators and the wave impedances of the phasing system was implemented using individual reactive four-terminal circuits with symmetrising dipole devices at their inputs. For this same purpose, high-sensitivity, interference-stable antenna amplifiers based on a multi-terminal circuit were used at the antenna outputs (before the system bringing about their mutual phasing). In the development and construction of the UTR-1, much attention was paid to accurate investigations of its main characteristics (the directive antenna gain, efficiency, effective area, antenna beam, side-lobe level etc.), taking into account all factors influencing the accuracy of measurements (interfaces, the ionosphere). Precision antenna measurements were carried out using a specially developed, automated apparatus and the radiation of strong cosmic radio sources and the background radiation, as well as computational and theoretical methods. Thanks to all these measures, the UTR-1 became an instrument that was suitable for absolute, accurate measurements of the intensities of cosmic radio sources. The maximum effective areas of the North–South and East–West antennas without allowing for losses (when the beam was pointed at the zenith and at the lowest frequency) were 1.5×10^4 and 1.2×10^4 m^2, respectively; the maximum resolution (when the beam was pointed at the zenith and at the highest frequency) was 35′ North–South and 80′ East–West.

After these instruments were built, each was used to investigate various questions associated with the construction of optimal, precision, broadband, decametre radio telescopes with electronically controlled beams, as well as for a wide range of radio astronomy programmes. The results of this work were published in more than a hundred articles and reviews; some of the results will be presented below. We emphasise that the series of theoretical and experimental (including technical) studies carried out by researchers at IRFE on radio astronomy antennas (Megn, L. G. Sodin, Yu. M. Bruk, N. K. Sharykin, P. A. Mel'yanovskii, G. A. Inyutin, N. Yu. Goncharov, V. P. Bovkun) and radio astronomy (L. L. Bazelyan, I. N. Zhuk, B. P. Ryabov) during the construction of the antennas and radio apparatus for the decametre radio telescopes, especially the UTR-2, led to a complex of new electronically controlled antennas. The principles of constructing such antennas proposed in these works can be used to built precision decametre instruments that are technically comparatively simple and reliable. The most characteristic properties of these antennas are their use of multi-element antenna grids consisting of broadband radiators with optimal configurations; an electronically controlled broadband beam in either one or two angular coordinates created using temporal phasing systems; isolated, multi-stage, parallel circuits for feeding and phasing the radiators with synchronous and asynchronous methods for pointing the beams; and discrete remote switching of the temporal delay lines, constructed according to the duality principle. The investigations carried out included theoretical calculations of the directive gains and efficiencies of large antennas near the semi-conducting interface (Sodin); development of a method for matching the frequency-dependent input radiator impedances in broadband, electronically controlled antennas in two angular coordinates and the wave impedances of the phasing system (Megn, Mel'yanovskii); theoretical analyses of electronic, asynchronous systems for beam control (Bruk, Sodin); and the creation of optimal, multi-stage distributive antenna-gain circuits based on interference-stable, multi-terminal amplifiers (Megn, Bovkun, K. A. Babenkov). In addition, principles for the accurate high-frequency control of large electronically controlled antennas were developed (Megn, Mel'yanovskii), and work was carried out on the creation of broadband baseline elements, discretely switchable temporal delay lines (Bruk, Inyutin, Mel'yanovskii, Sharykin, Goncharov) and new automated measurement technology (A. V. Antonov, Bovkun, Megn), as well as a number of other topics. The results of these studies have wide applicability in receiver and measurement technology and the technology of long-distance, short-wavelength links.

The most recently constructed decametre-wavelength telescope of IRFE—the UTR-2—was planned and erected from 1964–1969, was tested in 1970–1971, and began to be used for regular scientific observations in 1972. The technical work on the development of the UTR-2 (Braude, Megn, Sodin) foresaw the creation of a highly sensitive instrument that had a high directivity in two planes, a resolution (at its highest frequency) of about $30 \times 30'$ and an effective area (without taking into account losses) no lower than 10^5 m^2. The calculated sensitivity of the radio telescope (5–10 Jy) was expected to yield a volume of scientific information at decametre wavelengths for its main programs that was comparable to that obtained in the better studied metre- and decimetre-wavelength regimes.

Carrying out precision absolute measurements required referencing of the main instrumental characteristics in a broad field of view and over the full range of operational frequencies. Due to the influence of the refraction of radio waves in the

Fig. 8.4 North–South antenna of the UTR-2 radio telescope, comprised of 1440 oscillators

ionosphere in the case of highly directive antennas, the radio telescope was to have a multi-beam (in terms of zenith angle) antenna beam. The use of the UTR-2 for various scientific research required a rapid and operative change in the beam orientation in both planes, in accordance with a specified programme. The radio telescope was to be distinguished by high reliability and a simple construction.

Like its smaller cousin the UTR-1, the UTR-2 radio telescope was intended for operation at 10–25 MHz, and had an optimal T-shaped configuration that ensured matching of its resolution and sensitivity. The antenna systems, which occupy about 16 hectares, consist of two multi-element grids placed so as to enable regulation of the beam sidelobe level in both planes. The first antenna is oriented North–South, has a length of 1860 m and a width of 54 m and contains 1440 radiators forming 6 rows from North to South with 240 parallel dipoles in each (Fig. 8.4). The second antenna is oriented perpendicular to the first antenna (East–West) at its centre, has a length of 900 m and a width of 40 m and consists of 600 radiators placed in 6 rows from North to South with 100 co-linear dipoles in each row. The resulting pencil beam for the radio telescope as a whole is formed by multiplying the beams of the two individual antennas. The size and arrangement of the dipoles is the same as for the UTR-1. The maximum total effective area of the antennas (at 10 MHz with the beam pointing at the zenith) without taking into account losses is 152 000 m^2, and the minimum half-power beam width (at 25 MHz with the beam pointing at the zenith) is $30 \times 30'$.

The circuit for feeding the dipoles of both antennas is multi-stage and parallel with good broadband isolation of all the high-frequency tracts. The broadband input resistances of each dipole are optimally matched to the wave impedances of the feeding and phasing system, taking into account the dependences of the input resistances on the frequency and the two angular coordinates determining the pointing direction of the beam. The radio-telescope beam is pointed electronically in the sector from -30 to $+90$ in declination and from ± 3.5 hr to ± 12 hr (depending on the declination) in hour angle. This is brought about by five phasing systems: two are used in each of the antennas to point its beam in the two angular coordinates, and the fifth is used to mutually phase the antennas. All these systems are broadband, constructed according to the duality principle, and make use of temporal delay lines with discrete remote control. The two most complex phasing systems, which control the highly directed multipliers of the antenna beam, are based on a multistage, asynchronous circuit, while the remaining phasing systems use a single-stage, synchronous circuit. In the field of view indicated above, $2^{21} = 2\,097\,152$ beam positions can be used: 2^{11} in declination and 2^{10} in hour angle. The beam position can be changed from the control panel either by hand or by computer, ensuring virtually continuous tracking of a selected object over no less than 7 hours. To allow for refraction in the ionosphere and to ensure stability against interference and efficiency of observations, the resulting UTR-2 antenna beam is formed with five beams separated by half a beam width in declination, each having independent outputs. Observations can be conducted simultaneously on these five beams, and within each beam, at the six decametre frequencies 10, 12.6, 14.7, 16.7, 20 and 25 MHz. In this case, measurements can be taken simultaneously in 30 receiver channels, whose output signals are recorded on tape and then also sent to a computer, where the reduction is carried out in real time by a series of specialised programmes.

To realise the maximum sensitivity of the instrument while maintaining the required stability against interference, compensation for losses in the UTR-2 antennas is performed by a two-stage antenna-amplification system consisting of 21 broadband amplifiers developed specially for this purpose using an optimal multi-band circuit. The UTR-2 beam shape can be adjusted to decrease the effect of confusion, making it possible to decrease the sidelobe level in declination by the introduction of the necessary current distribution, and the sidelobe level in hour angle through appropriate processing of the data.

The good condition of the numerous radio devices incorporated into the UTR-2 is tested using two specialised high-frequency control systems, enabling the operational remote verification of the condition of the antenna networks, and of the entire antenna and receiving apparatus of the radio telescope.

The efficiency and reliability of the UTR-2 have been verified over a decade of nearly continuous use of the telescope. This instrument remains unique to this day in terms of its design, characteristics and abilities, and is fundamentally different from other instruments operating in this wavelength range. However, the resolution provided by the UTR-2 is insufficient for a number of scientific tasks, such as studies of the angular structure of cosmic radio sources. These and other such measurements require a resolution at decametre wavelengths of the order of several arcseconds. Although such resolutions can be provided under specific conditions, such as analysis of scintillations of the intensities of radio sources due to inhomogeneities in the interplanetary medium or observations of the occultation of a source by the disk of the Moon (indeed, such observations are carried out on the UTR-2), the most effective means to obtain this higher angular resolution is by using an array of radio interferometers with sufficiently long baselines. With this goal in mind, the Radio Astronomy Division of IRFE began the development and construction of the URAN system in 1973. The main element of this system was the UTR-2, which forms an interferometer array together with four relatively small radio telescopes, yielding a maximum resolution of about 1 arcsecond. The first interferometer of this system is URAN-1, with a baseline of 42.6 km and a resolution of about 30 arcseconds, and is comprised of the North–South UTR-2 antenna and a second URAN-1 antenna located near the town of Gotbal'd. This instrument was already completed by the end of 1975, after which test observations began. The Odessa Division of the Main Astronomical Observatory of the Academy of Sciences of the Ukrainian SSR (scientific supervisor V. P. Tsesevich) and the Poltav Gravimetric Observatory (scientific supervisor V. G. Bulatsen), in collaboration with IRFE, began the development and construction of the URAN-4 antenna near the village of Mayak in the Odessa region in 1975 and the URAN-2 antenna near Poltav in 1979. These radio interferometers have baselines of 592 and 176 km, and should provide resolutions of about two and about seven arcseconds. The maximum resolution (of about one arcsecond) will be provided by the baseline to the URAN-3 antenna, which will be located near Lvov.

The specialised antenna systems of all the URAN radio interferometers are designed to operate from 12.6 to 25 MHz, and are each comprised of a multi-element, equidistant, rectangular array with an electronically controlled beam in both coordinates. The arrays consist of 128 or 256 radiators. To enable elimination of the influence on the interference measurements of the rotation of the plane of polarisation of linearly polarised radio waves in the ionosphere (ionospheric Faraday rotation), which is very substantial at decametre wavelengths, all the URAN antennas are equipped to receive simultaneously two linear or circular polarisations. For this purpose, tourniquet radiators are used in the URAN antenna arrays, each consisting of two orthogonal, linear, horizontal, broadband dipoles. The dipoles for each linear polarisation have their own broadband feed and phasing systems, based on a multi-stage, synchronous circuit. These phasing systems are controlled in parallel, so that each antenna is made up of two superposed multi-element arrays for the reception of linearly polarised waves, with separate outputs. Photographs of the dipoles of the URAN-1 antenna and the entire antenna array have been taken from an aircraft (Fig. 8.5).

The URAN-1 radio interferometer is intended for interferometric observations in real time, while the remain URAN interferometers are designed for use as a Very Long Baseline Interferometry system.

8.3 Radio Astronomy Studies

As was already noted above, a variety of radio astronomy observations were carried out on all the decametre instruments of the Radio Astronomy Division of IRFE beginning in 1960. Most recently, an especially wide range of observations have

Fig. 8.5 Antenna grid of the URAN-1 system

been conducted on the UTR-2 (from 1972) and the URAN-1 interferometer (from 1975). Of all the decametre-wavelength scientific programmes carried out over these years, the main ones are

(1) measurements of the flux densities of discrete radio sources and compilation of a catalogue of these values;
(2) determination of the angular structures of radio sources (radio imaging) using various methods;
(3) observations of the radio emission of pulsars;
(4) studies of the Galactic background, in particular, of regions of ionised hydrogen (H II regions);
(5) radio spectroscopic observations and searches for various radio lines;
(6) studies of the radio emission of the Sun, as well as of the scattering of radio waves in the solar corona;
(7) searches for and studies of non-thermal emission from planets in the solar system.

We will now briefly discuss some results of these studies.

Among the various objects that are studied in radio astronomy, discrete radio sources are of special interest. About 30,000 such sources are currently known, and this number will, of course, grow with the sensitivity and resolution of radio telescopes. Some of these sources have already been identified with either Galactic (supernovae, gaseous nebulae etc.) or extragalactic (normal, radio, Seyfert and N-type galaxies, quasars etc.) objects. These sources had been studied in the radio primarily at metre through centimetre wavelengths, while observations at decametre wavelengths were much rarer. It was therefore of considerable interest to compile a catalogue of discrete radio sources at decametre wavelengths. Such a catalogue was created using the UTR-2 at 5 to 6 wavelengths from 12 to 30 m. More than 1500 sources have currently been investigated in the area of sky with declinations from $-13°$ to $+20°$ and from $+50°$ to $+60°$, 60 of them discovered for the first time.

Although some of these sources are thermal, the vast majority are non-thermal, especially the extragalactic sources. The main mechanisms giving rise to this non-thermal emission are Bremsstrahlung and the synchrotron mechanism. This is testified to by the frequency dependence of the emission intensity, which usually falls off with increasing frequency in accordance with a power law. In the case of thermal mechanisms, the radio intensity first grows with frequency in proportional to the square of the frequency, then becomes nearly constant at very high radio frequencies. If we plot the spectrum of a non-thermal source—the dependence of the spectral flux density on the frequency—on a logarithmic scale, this dependence is linear, with the slope, called the spectral index (α), being different for different discrete sources. It turns out that the energy spectrum of relativistic electrons whose motion in the source magnetic fields gives rise to the radio emission is also a power law, with index γ. According to the theory of synchrotron radiation, these two indices are related by the simple expression $\gamma = 2\alpha + 1$.

If we proceed from this position, which has been confirmed by observations over a wide range of wavelengths (from centimetres to metres), we can use various theoretical models to predict the frequency spectra of non-thermal sources at decametre wavelengths. In general, we would expect spectra of two types: linear if there are no factors altering the source spectrum, and curved, with negative curvature of the spectrum at decametre wavelengths. The latter type of spectrum should be observed, for example, if there is self-absorption in the source, absorption due to H(II) or some other similar mechanism. Theoretically, there should be many sources displaying self-absorption, since measurements at higher frequencies show quite a few sources where this effect is clearly visible. Observations carried out in the USA and Canada first seemed to support this picture.

A series of investigations were carried out at IRFE, whose goal was to measure the flux densities of discrete sources in order to construct their spectra at frequencies 10–25 MHz (Braude, Zhuk, Ryabov, K. P. Sokolov, Sharykin and others). There are currently data enabling the construction of spectra at frequencies 10–1400 MHz for about 700 sources. The number of such sources will be increased to 1000 in the near future. It turns out that, in addition to the expected two types of spectra described above, two more qualitatively different types of spectra are also observed. The data showed both linear (I) spectra and curved spectra with negative (II), positive (III) and sign-variable (IV) curvature. The relative numbers of sources in these categories are

Type of spectrum	I	II	III	IV
Number of spectra, %	86	3.5	10	0.5

As we can see, the vast majority of sources in this frequency range have linear spectra (type I) and rather few have type II spectra, in contrast to the expectations based on higher-frequency data. A fairly large group of sources has type III spectra, which were not previously observed (with an increase in the spectral flux density with decreasing frequency more rapid than linear behaviour).

To elucidate how discrete sources with different types of spectra differ from each other, the spatial distribution of radio galaxies, quasars, supernova remnants and unidentified objects with type I, II and III spectra were compared. It turns out it is not possible to establish a connection between type of spectra and either spatial location or type of source. The same is true of other properties that were investigated.

This led to the suggestion that, independent of the type of radio source, there exists some common factor leading to these various types of spectra. It was shown (Braude, Zhuk, Megn, Ryabov) that one possible such factor was a deformation of the energy spectrum of cosmic-ray electrons with energies of 10^7–10^8 electron volts. It is precisely at these energies, in magnetic fields of 1–10 milliGauss, that decametre radio waves are generated in discrete sources.

Thus, based on our observations, it was possible to show that, contrary to theoretical predictions, sources with type II spectra are encountered only rarely, while quite a few sources with type III spectra are observed.

A very important conclusion flows from these studies: the vast majority of sources that high-frequency observations show to have features with small angular sizes should show signs of self-absorption. However, as our data show, self-absorption is not observed in most of these sources, right to decametre wavelengths. One possible explanation for this is that decametre radio waves are intensely scattered on inhomogeneities in the plasma in the sources themselves. This scattering could lead to an appreciable increase in the angular size of a source, and so sharply weaken the effect of self-absorption. It is also possible that these sources have extended features or a halo whose spectra are characterised by large spectral indices, and that these features make a substantial contribution to the total radio flux at decametre wavelengths. In this case, we should also observe a corresponding change in the radio images with decreasing frequency. This leads to the need for independent measurements of the angular structures of various types of radio sources at decametre wavelengths.

Four different methods for resolving extended radio sources are being used at IRFE to attack this problem: UTR-2 observations with a narrow antenna beam; analysis of scintillation of the intensity from radio sources on inhomogeneities in the interplanetary plasma; observations of radio sources occulted by the disc of the Moon; and interferometric observations. All of these methods have been intensely applied under the scientific supervision of Braude and Megn, and information about decametre-wavelength angular structure is already available for a number of objects (Vovkun, Zhuk, A. L. Bobeiko, Yu. Yu. Sergienko and, in recent years, S. L. Rashkovskii, I. S. Fal'kovich, Sharykin, V. A. Shepelev and A. D. Khristenko).

For example, results obtained on the UTR-2 using the last three methods have been used to study the decametre angular structure of the important radio source the Crab Nebula. Strip radio brightness distributions at several frequencies have been determined from these observations. The coordinates of the compact source in the Crab Nebula have been measured for the first time at decametre wavelengths, and coincide within the errors with the position of the pulsar NP0532. The relative contributions of the compact source to the total radio flux, measured with very high accuracies for the decametre range, are 0.32 ± 0.03, 0.42 ± 0.04 and 0.55 ± 0.08 at 25, 30 and 16.7 MHz, respectively. This source has a linear spectrum (on a logarithmic scale) at 16.7–122 MHz, with spectral index $\alpha = 2.09 \pm 0.04$. A flattening and turnover of the spectrum at low frequencies is observed for the spectrum of the Crab Nebula without its compact source.

Observations of the quasar 3C196 on the URAN-1 interferometer together with measurements of scintillation of its intensity were used to identify the best model from a number of models that had been considered earlier. At decimetre and centimetre wavelengths, this quasar has two features with angular sizes of no more than $2 \times 2''$ separated by about $5''$ on the sky, while it displays an extended region of radio emission at decametre wavelengths, with the intensities of compact features and extended emission being nearly equal. The extended emission in 3C196 spans about $18''$ in right ascension and $25''$ in declination.

Similar measurements aimed at establishing good models for radio sources for which new and interesting data were obtained have also been made for the supernova remnant Cassiopeia A, the radio galaxy 3C280 and a number of other cosmic radio sources.

Pulsar studies on the UTR-2 began in 1972 (Bruk, B. Yu. Ustimenko). For a number of reasons, it had been considered very doubtful that useful observations of pulsars could be carried out at decametre wavelengths. Nevertheless, using original instruments, it was possible to detect for the first time pulsed structures in the signals for 7 of 12 nearby pulsars observed at 10 to 25 MHz. It was found that the decametre radiation of pulsars is very different from their radiation at higher frequencies. In particular, a larger number of interpulses are observed at low frequencies, which are virtually completely absent at high frequencies. The spectra of the main pulse and interpulses are different, as well as the character of their time variations.

The interpulses arise sporadically, but their location between two neighbouring main pulses is fairly fixed. Interpulses were observed for all pulsars that were detected at decametre wavelengths, with the intensities of the main pulse and interpulses being comparable. Moreover, it was found that the intensity of the main pulse falls with decreasing frequency in the range 16.7–25 MHz, while the intensity of the interpulse grows.

Simultaneous observations of 5 pulsars were obtained over a wide range of frequencies, from 16 to 1400 MHz. These studies were carried out at the Grakovo (Ukrainian SSR), Pushchino and Jodrell Bank (Great Britain) observatories. The frequency spectra of the main pulses and interpulses were measured. The main pulses of the five studied pulsars had their maxima at frequencies of 80–120 MHz, while no such maxima were observed for the interpulses.

The structure of the decametre radiation of pulses is very complex. In addition to the main pulse and interpulses (up to 5 can be observed, with amplitudes form 20 to 100% of that of the main pulse), an extended component was observed, as well as flares of the interpulse emission lasting 30–60 minutes, whose amplitudes and energies could sometimes appreciably exceed those of the main pulse. Although the amplitude fluctuations for the interpulses and the variability of their structures are substantially greater than for the main pulse, the mean profile of the signal proves to

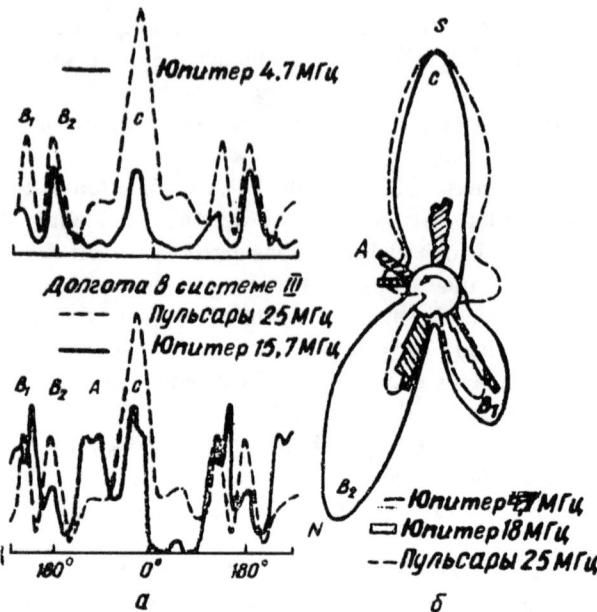

Fig. 8.6 Comparison of the longitude structure of the emission of pulsars and Jupiter. The *upper left text* reads "Jupiter 4.7 MHz," the *middle left text* "Longitude in the system," "Pulsars 25 MHz," and "Jupiter 15.7 MHz," *from top to bottom*, and the *lower right text* "Jupiter 4.7 MHz," "Jupiter 18 MHz," and "Pulsars 25 MHz," *from top to bottom*

be very stable. For many pulsars, it is possible to identify structural elements in the received signals that repeat often and are stable. A comparison of these structures at decametre and shorter wavelengths suggested the presence of common characteristics of all pulsars. As was noted above, there are symmetrically placed zones in the pulsar period where the signal is received.

The presence of such structures suggested there might be a similarity between the decametre radiation of pulsars and of Jupiter (Fig. 8.6). The decametre radiation of Jupiter is manifest as intense, sporadic outbursts with various characters and origins. Above 20 MHz, the structure of the radiation is subject to strong influence due to Jupiter's moon Io.

The longitude structure of the radiation of pulsars and Jupiter was compared. The influence of Io was excluded using longitude profiles of Jupiter at frequencies below 20 MHz. Supposing that the radio emission of pulsars originates at their magnetic poles, the emission regions for Jupiter and pulsars were compared after shifting the centre of the main pulsar pulses to their South poles. The results show that the structures of the impulsive emission of these two completely different types of astronomical bodies are quite similar. This resemblance could be associated with similar configurations for the magnetic fields of these objects (Braude, Bruk). Jupiter has a close to dipolar magnetic field, with the dipole inlined by 10° to the rotational axis of the planet and shifted relative to the centre of gravity. In some theories, the radio emission of pulsar is also explained by the presence of a dipolar magnetic field that is inclined and shifted relative to the centre of the object. Of course, the strengths of the magnetic fields, the inclinations of these fields to the rotation axis and a host of other physical properties are very different for pulsars and Jupiter, but the similarity in the emission structure provides hope that further detailed studies of the radio emission of Jupiter will help bring us closer to an understanding of the physical processes giving rise to the radio emission of pulsars as well.

Stars can form in regions of ionised hydrogen (H II regions). The radio emission of these regions have been studied mainly using so-called recombination lines. Such studies have been carried out primarily at centimetre and decimetre wavelengths. This emission arises fairly deep inside the H II regions, and the derived information characterises precisely the inner parts of the region of ionised hydrogen (see Essay 1).

Observations of H II regions with the UTR-2 radio telescope, which has relatively high angular resolution, open new possibilities for studies of these regions. Some of the background radiation is absorbed as it passes through such regions. Observations of this absorption at decametre wavelengths can be used to determine a number of important parameters of the H II regions, and also to study the non-thermal radiation of the Galaxy. Since the optical depth of an H II region is rather high at decametre wavelengths, the main role in such absorption studies is played by the outer parts of the H II region. Decametre measurements supplementing measurements of recombination lines enable more complete analysis of the structure of such regions of ionised hydrogen.

A specialised method for measurements and calibration on the UTR-2 was developed for such observations (V. V. Krymkin). H II regions were observed using the five-beam UTR-2 antenna beam in a transit mode, with the studied regions passing through the beams due to the rotation of the Earth. Measurements were carried out at 5 frequencies from 12.6–25 MHz. The data obtained were used to estimate the electron density, n_e, and electron temperature, T_e, for a number of H II regions, as well as quantities characterising the distribution of non-thermal Galactic emission. Six H II regions were studied using the UTR-2 images obtained, including NGC1499, for which the values $T_e = 4400$ K and $N_e = 9$ cm^{-3} were derived. These values are in good agreement with measurements carried out at other frequencies, testifying that the outer and inner structures of NGC1499 are similar. The results of such observations of various regions were used to estimate the mean density of non-thermal radio emission along the lines of sight from the Earth to these objects. These estimates coincide with previously determined values, and confirm that there is a higher density of radiation in the arms of the Galaxy than in the inter-arm regions. Thus, observations at decametre wavelengths can provide independent information about the temperatures and densities of H II regions, thereby verifying various conclusions about these structures based on observations at shorter wavelengths.

The methods developed for studies of H II regions were also used to investigate old supernova remnants. One example is the detection and explanation of the low-frequency turnover in the spectrum of a source in the Monocerous Loop. It was concluded that this turnover is due to the presence of ionised material, which directly absorbs the source radiation, as well as the background emission generated along the line of sight toward the source. The resulting drop in the brightness temperature of the background is balanced by the remaining source emission, giving rise to an apparent turnover in the spectrum.

Until recently, studies in the intensely developing area of radio spectroscopy were carried out from millimetre to metre wavelengths. The construction of the UTR-2 radio telescope made possible radio spectroscopic observations at decametre wavelengths. Through such studies (A. A. Konavalenko, Sodin) indicated the possibility of detecting recombination lines of hydrogen with very high principle quantum numbers $n = 630$–650, whose wavelengths are in the decametre range. A special spectrometer with a frequency resolution of 1 kHz and a sensitivity of $\Delta T/T \simeq 4 \times 10^{-4}$ for an integration time of 10 hours was developed for such observations. However, the numerous observations carried out in various areas of the sky did not give positive results, and no such hydrogen recombination lines were detected.

This same equipment was used to attempt to detect a line of neutral nitrogen, ^{14}N. Back in the 1950s, I. S. Shklovskii indicated the existence of two lines of ^{14}N at decametre wavelengths, with frequencies $f_2 = 26.127$ MHz and $f_2 = 15.676$ MHz. Calculations showed that these lines should be observed in absorption with very small optical depths, $10^{-4} - 10^{-5}$. As we can see, these theoretical calculations were not very encouraging; nevertheless, observations were conducted on the UTR-2 with the aim of attempting to detect the presence of nitrogen in space (Sodin, Konavalenko). The experiment was set up to try to detect nitrogen in our Galaxy. It is known from the literature that strong absorption in the 1420-MHz line of neutral hydrogen HI is observed in the direction of the strong radio source Cassiopeia A, which occurs in the spirals arms of our Galaxy—in particular, the so-called Orion and Perseus arms. It was logical to look for nitrogen absorption in this same direction. Thus, the observations made use of a spectrometer, and were carried out on the UTR-2 in the direction of Cass A. Absorption was detected close to 26 MHz, both in the nearer Orion arm and the more distant Perseus arm. The radial velocities in the two arms were different, with the velocity in the Orion arm being about 42 km/s and that in the Perseus arm being near zero. This line was found to be fairly intense for the Perseus arm, but weak for the Orion arm. The Galactic origin of the detected line is confirmed not only by the coincidence of the radial velocities for this line and for the HI absorption in this direction, but also by the presence of periodic variations in the frequency of the line, which coincide with the Doppler shift due to the revolution of the Earth around the Sun. If the detected line is taken to be the sought-for ^{14}N line, the data obtained can be used to determine the nitrogen abundance in these two arms in our Galaxy. In the case of the Perseus arm, the observed optical depth of the line turned out to be 2×10^{-3}, appreciably higher than predicted by the theoretical calculations. Assuming that the spin temperatures for the HI and ^{14}N are the same, for the observed ratio of the densities of HI and nitrogen, the abundance of ^{14}N to HI is about 130–140, instead of the commonly adopted value of 2000.

This discrepancy led American radio astronomers to suggest that the absorption line observed on the UTR-2 was a recombination line of carbon with principle quantum number $n = 631$. They pointed out that this line, in contrast to all recombination lines known at the time, should be observed in absorption rather than emission. If this was the observed line, all the observational data would be in agreement with generally accepted abundances.

Experiments carried out at IRFE in 1981–1982 fully confirmed this point of view. It was possible to push the measurements toward longer wavelengths, and recombination lines of carbon corresponding to transitions from levels with principle quantum numbers up to $n = 732$ were detected.

The solar corona is another very interesting object. Measurements at decametre wavelengths can be used to study high layers in the corona, out to two solar radii. This region is of considerable interest, since it is here that the solar wind is generated, and it is currently poorly studied. Thanks to its large effective area (sensitivity), broadband receivers and antenna beam adjustable in two coordinates, the UTR-2 could be used to derive the two-dimensional distribution of the radio brightness of the Sun. This telescope was thus used to carry out detailed studies of the Sun during partial eclipses.

Studies of the solar corona at distances from 5 to 25 solar radii were conducted by analysing the radiation of discrete background sources that had passed through the solar plasma, and deriving the radio images of these sources distorted by scattering and refraction in the corona. It was found that the increase in the effective sizes of sources inversely proportional to the square of the wavelength observed at metre and shorter wavelength slows at decametre wavelengths. This made it possible to explain some properties of scattering of radio waves on inhomogeneities in the solar plasma near caustics (Bazelyan, Braude, Megn, V. G. Sinitsyn). The observational data that were obtained were used to refine certain parameters of the solar corona.

A large amount of time in the UTR-2 observing programme was dedicated to studying the sporadic component of the solar emission. Together with flares with durations from several seconds to several hours, the decametre observations revealed short-lived flares with lifetimes from several seconds to tenths of a second. These were new manifestations of type III outbursts characteristic only of the decametre range. These flares are elementary events, and are often subdivided in frequency or time, forming chains. In a number of cases, these phenomena were observed to precede ordinary type III outbursts. These observations were carried out on the UTR-2 in collaboration with radio astronomers of IRFE and the Radio Physics Research Institute in Gorkii (Balezyan, E. P. Abranin, Zaitsev, V. A. Zinichev and others).

As is noted above, the theoretical possibility of radar studies of the solar corona was investigated at IRFE as early as 1957; the required power was estimated and an optimal frequency range was selected. These calculations were confirmed by the successful realisation of radar sounding measurements at 25 and 38 MHz in the USA, however the new results were never adequately explained. In 1968–1974, a series of studies was carried out at IRFE (I. M. Gordon, N. N. Gerasimova) dedicated to creating a new theory for the formation of radar signals reflected from the solar corona.

It was shown that all important features in the structure of the reflected signal could be explained if it was supposed that the signal was formed and amplified in turbulent zones of the corona by induced scattering on turbulent pulsations of the coronal plasma. It turned out that the signal was reflected only from regions of the corona above active regions, whereas they were absorbed by the "quiet" corona.

Attempts to use the UTR-2 to detect decametre radiation of planets in the solar system were also made. Such radiation was detected only from Jupiter, which was known to be a powerful source of decametre radio emission. The fine structure of this emission is currently being studied.

Together with experimental work, a large amount of attention has also been given to theoretical studies, which encompass a wide range of questions, from the processes occurring in discrete sources to phenomena in the solar corona (V. M. Kontorovich, A. V. Kats, A. E. Kochanov, Ya. M. Sobolev, Yu. A. Sinitsyn, N. A. Stepanova, V. N. Mel'nik, V. I. Vigdorchik and others).

It is known that astrophysical objects are rarely in a complete state of thermodynamical equilibrium, but nevertheless, many of these objects display nearly universal power-law spectra. For example, power-law energy (frequency) distributions are characteristic of many cosmic-ray sources, for cosmic rays themselves and, in a number of cases, for the optical, X-ray and sometimes gamma-ray emission of sources.

The theoretical studies carried out established a general mechanism for the appearance of power-law spectra, which are characteristic for Kolmogorov turbulence. It was shown to be of considerable interest to study possible ways to form Kolmogorov spectra in astrophysical sources. A number of general properties of power-law spectra in weakly turbulent systems of waves and particles were elucidated. The relationships between equilibrium and power-law spectra were studied, and the corresponding power-law distributions in a strong magnetic field derived. Non-linear spectra obtained due to synchrotron or Compton losses in the case when there is a localised injection of hard particles (in other words, when the source of relativistic electrons is a localised region of high energy) were also studied. It was shown that, in an inhomogeneous source under these conditions, the radiation flux that diffuses outward forms a non-linear spectrum with linear power-law sections, including sections with the "universal" power-law with spectral index equal to two. At the same time, spectra with both positive and negative curvature could be formed. The role of inhomogeneity of the magnetic field and diffusion in "core + halo" sources was studied, and it was shown that the spectra of galaxies with low-frequency turnovers whose rises were not too steep could be explained in this scenario.

In addition to questions associated with the formation of power-law spectra, a number of problems connected with the physics of the solar-corona plasma were studied. In particular, the propagation of electron beams in the corona was analysed, taking into account the inhomogeneity of the plasma. It was shown that a beam of electrons ejected from the region of a solar flare propagates toward regions of lower plasma density. As a result, plasma waves "lag" behind the beam, and come out of resonance with it. Due to this movement of ejected clumps of plasma from resonance in the inhomogeneous corona, the characteristic relaxation length for the beam grows to scales of the order of the dimensions of the corona. This stabilises the beams observed during type III outbursts.

Theoretical studies also examined the role of non-equilibrium conditions in plasma in the solar interior arising due to the nuclear reactions occurring there. It was shown that such conditions are associated with the existence of power-law energy "tails" in the velocity distribution for the elements; due to these high-velocity "tails," the rates for nuclear reactions involving elements with high Coulomb barriers turn out to be higher than for a purely Maxwellian velocity distribution. Hydrodynamical stability and interactions of surface and internal waves with turbulence was also studied, as well as wind instability in the lobes of radio galaxies.

In this essay, we have briefly examined the results of studies carried out from 1959 through 1980. In 1980, the Radio Astronomy Division was organised in the Institute of Radio Physics and Electronics, by a resolution of the Presidium of the Academy of Sciences of the Ukrainian SSR (L. N. Litninenko was appointed chairman of the Division). In this Division, radio astronomy studies at millimetre wavelengths were begun, alongside continued work at decametre wavelengths. A quantum amplifier operating at 45 GHz that has increased the sensitivity of radiometers in this range by an order of magnitude (to $\Delta T = 0.05$ K) has already been developed and installed on the RT 25-2 radio telescope of the Institute of Applied Physics of the Academy of Sciences.

Chapter 9
Radio Physical Studies of Planets and the Earth at the Institute of Radio Technology and Electronics of the USSR Academy of Sciences

B.G. Kutuza and O.N. Rzhiga

Abstract Stages in the development of radar astronomy and the main methods and scientific results obtained at the Institute of Radio Technology and Electronics (IRE) beginning in the 1960s are discussed. The fundamental importance for manned space flight of the radical improvement in the accuracy of the astronomical unit achieved using radar measurements of planets is underscored, together with other important results of studies of planetary surfaces. Studies of the atmosphere and surface of the Earth using radio-astronomy methods are described in detail, including satellite techniques. The results of radio-physical studies of the ionospheres of the Moon and planets carried out by analysing signals transmitted by spacecraft and received on the Earth are also presented.

9.1 Radar Astronomy[1]

In the middle of 1960, at the initiative of Academician V. A. Kotel'nikov, the Director of the Institute of Radio Technology and Electronics of the Academy of Sciences of the USSR [IRE], preparations were begun for radar measurements of Venus, which is the planet that approaches the closest to the Earth. Since that time, a series of work on radar studies of planets, led by Kotel'nikov, has been one of the main research directions at the institute.

In the first experiments on radar measurements of the Moon carried out in 1946 in the USA and Hungary, radar stations intended for the detection of airplanes were used. To detect signals reflected from Venus, the energy potential (sensitivity) of the radar installations used had to be increased by a factor of 10 million, and then increased by a further factor of 100 to detect reflected signals from Mars. Thus, signals reflected from other planets could be detected only after considerable development of antenna, transmitter and receiver technology, and the development of methods for signal detection.

---

[1] Section 9.1 was written by O. N. Rzhiga and Sect. 9.2 by B. G. Kutuza.

Fig. 9.1 Antenna of the Deep Space Communication Center in the Crimea, consisting of eight parabolic reflectors 16 m in diameter on a common azimuthal–vertical steerable structure

Creation of a Radar Installation for Planetary Studies Yu. K. Khodarev suggested to use the antenna and transmitter of the Deep-Space Communications Centre (DSCC) that were then being built near Evpatoria for radar measurements of Venus. This antenna was being constructed under the supervision of E. B. Kopenberg, and the transmitter of continuum radiation at a wavelength of 39 cm under the supervision of V. P. Minashin (Fig. 9.1). The base of a rotating gun turret from a linear ship was used as the mount platform (see Essay 1). The antenna, intended for flight control and the reception of information from interplanetary stations, was designed in three months and then constructed in another three months.

In the middle of 1960, O. N. Rzhiga wrote that it would be possible to receive a reflected signal from Venus using the technical facilities of the DSCC during the nearest inferior conjunction in April 1961. It would thus be possible to measure the distance between Venus and the Earth using a signal sent from the DSCC transmitter. He devised a structural scheme for the radar installation and a method for the radar measurements, and also estimated various factors affecting their accuracy.

In view of the uncertainty in various parameters, it was proposed to record the reflected signal on magnetic tape, making it possible to select the optimal parameters for the measuring apparatus when playing back the recording. With this aim, V. M. Dubrovin developed a magnetic-recording system using a reference oscillator to ensure precise time marks when playing the recordings. Dubrovin developed a receiver with a five-fold frequency transformer. The ultra-high-frequency (UHF) part of the receiver, which had a parametric input amplifier, was developed under the supervision of N. N. Nikitskii.

The detection of the weak reflected signal against the fluctuating noise background of the receiver required a long integration time, up to several or even tens of hours, in the case of a very weak signal. The detection of the reflected signal and measurement of its frequency and energy at the receiver output were realised using a multi-channel spectrum analyser designed by V. A. Morozov with the help of E. G. Trunova. The energy of the signal was measured using the fact that the total time when the upper envelope of a certain threshold level is exceeded in each frequency channel of the analyser grows proportional to the energy of the signal. V. I. Bunimovich and Morozov showed that, under the relevant conditions, this method for the reception of a weak signal was essentially the optimal one. The spectrum analyser was constructed under the supervision of G. A. Podoprigory, and had 10 frequency channels with a transmission bandwidth of 20 Hz.

The radiated signal was a periodic succession of pulses and pauses with equal durations, which were used for the radiometer reception of the reflected signal and determination of corrections to the expected signal arrival time. The tie of the signal to Universal Time and determination of the expected signal arrival time were realised using a program-timing device devised by G. M. Petrov. Petrov also developed a device designed to switch off the transmitter before the onset of reception, switch the antenna from the transmitter to the receiver and change the polarisation of the antenna. Rzhiga and L. V. Apraksin developed a generator that could be used to vary the frequency of the transmitter carrier signal in accordance with the expected Doppler shift, in order to ensure reception of the reflected signal near the nominal frequency.

Calculation of the expected (ephemerides) values of the delay time for the reflected signal and Doppler shifts to the transmitter carrier frequency, as well as commands for pointing of the antenna, were carried out using electronic computers. The programming of these computers and conduction of the required computations was supervised by M. D. Kislik; participants in this work included D. M. Tsvetkov, B. A. Stepanov, B. A. Dubinskii and G. A. Zhurkina. The computations used tables of heliocentric positions and velocities of the centres of mass of Venus and of the Earth–Moon system, as well as tables of geocentric positions of the centre of mass of the Moon, compiled at the Institute of Theoretical Astronomy of the USSR Academy of Sciences by D. K. Kulikov and N. S. Subbotin. The computational programs took into account the non-simultaneity of the times for the radiation, reflection and reception of the signals and changes in the configuration of the planets in space over the elapsed time.

The DSCC antenna, which was intended for communicating with satellites and was set up to work with circularly-polarised signals, had to be equipped to receive the mirror-reflected signal from the planet. Kopenberg suggested the introduction of depolarisers in the form of linear grids in the tract to the receiver, between the feed and surface of the antenna, since this was simpler than changing the direction of rotation of the polariser plane between signal transmission and reception.

There was the possibility that the onset of the radar observations was delayed by delays in the manufacture of the depolarisers, antenna switch and parametric amplifier box. In February 1961, after the launch of the *Venera-1* automated interplanetary station, Academicians S. P. Korolev and M. V. Keldysh came to the DSCC complex. After Dubrovin and Rzhiga had reported on the state of the planetary-radar project, Korolev, who was interested in ensuring a successful flight of the interplanetary station, decreed that the work be urgently completed. The manufacture of the antenna

Fig. 9.2 Photograph done not long before the first radar observations of Venus. *Left to right*: A. M. Shakhovskoi, Academician V. A. Kotel'nikov, O. N. Rzhiga and V. M. Dubrovin

equipment was completed in the plant in a single night, and was sent to Evpatoria by fast train.

At the beginning of the observations, the radiated power was 10 kilowatts. During the observations themselves in the middle of April 1961, under the supervision of Minashin, four klystron transmitters were put together, increasing the radiated power by a factor of four, which played a decisive role in the success of these first observations.

The radar installation for planetary studies was established in the six months until the next inferior conjunction with Venus. This work was made possible by the experience and enthusiasm of the entire group of researchers involved. Some, from the Institute of Radio Electronics of the USSR Academy of Sciences, had experience in constructing receivers and regulating devices for observations of the first artificial Earth satellites and of automated stations sent to the Moon at the end of the 1950s; others had worked on the problem of distinguishing weak signals from noise. They were joined by a large number of staff from the leading organisations involved. A leading role in composing the group and organising the radar observations was played by A. M. Shakhovskii (Fig. 9.2). Much work in manufacturing, adjusting and servicing the equipment developed at the Institute of Radio Electronics was contributed by Yu. A. Alekseev, L. V. Apraksin, Yu. P. Vasil'ev, V. O. Voitov, A. V. Grigor'ev, O. K. Dmitriev, N. M. Zaitsev, A. V. Kaledin, P. P. Korsakov, B. I. Kuznetsov, P. V. Kuznetsov, E. F. Kushchenko, I. V. Lishin, Yu. M. Lobachev, L. P. Lundina, Yu. N. Paukov, G. A. Podoprigora, V. Kh. Sinitsa, N. M. Sinodkin, G. I. Slobodenyuk, O. S. Stepanov, A. L. Tamarin, B. K. Chechulin, A. S. Chikin and Yu. V. Filin. A large amount of help in organising and conducting the first observations of Venus was given by P. A. Agadzhanov, A. P. Rabotyagov and G. A. Sytsko.

Academician Kotel'nikov directly oversaw the creation of a unique radar installation at the DSCC, as well as the preparation for and conducting of the observations. The participants in the first observations remember working 16 hours or more in a day in order not to miss the optimal period for the Venus observations. The inexhaustible energy and optimism of Kotel'nikov supported a cheerful mood in the group even at the most difficult moments.

First Observations of Venus The first signal to Venus was sent on April 1, 1961. The main observations were carried out on April 18–26, 1961. At that time, the distance between the Earth and Venus was 43.5–47.5 million km. The observations were carried out at a wavelength of 39 cm, with circularly polarised radiation. A total energy of approximately 15 Watts was incident on the visible surface of Venus over a time of about five minutes. Approximately 20 s before the expected time of arrival of the reflected signal, the transmitter was switched off to that it would not contribute interference, the antenna was switched over to the receiver, the polarisation of the antenna was changed to linear and the reflected signal was received for five minutes.

The reflected signal was not detected in these first observations. This led to the suggestion that the spectrum of the reflected signal was smeared over a broader band than had been supposed ($\sim$20 Hz); the widths of the frequency channels in the spectrum analyser were accordingly expanded to 60 Hz. The real reason for the absence of a reflected signal was discovered on April 18, soon after the power of the transmitter had been increased fourfold. It turned out that one of the frequency dividers had been adjusted incorrectly, so that the changes in the pulses and pauses in the radiation, as well as the switching of the receiver apparatus to a radiometer regime, were essentially occurring at random times, hindering accumulation of the signal.

During the main observations (after re-adjusting the frequency divider), the repetition period for the pulses and pauses in transmission was either 0.256 s or 0.128 s, to enable unambiguous determination of the distance to Venus. Of the various possible values for the astronomical unit calculated based on the measurements of the distance to Venus, the value initially chosen was that closest to the value derived earlier using astronomical methods, which proved to be incorrect.

Immediately after the end of the observations, an analysis was carried out of the magnetic recordings of the reflected signal, which were obtained using electromechanical filters with a transmission bandwidth of 4 Hz placed in the channels of the spectrum analyser. The use of these filters, developed by M. G. Golubtsov, made it possible to remove ambiguity in the distance measured from the Doppler shift of the central frequency and correctly determine the width of the received spectrum.

In the mean spectrum of the signal reflected from Venus on April 18, 1961 obtained by analysing five receiver sessions of about five minutes each (Fig. 9.3), the width of the spectrum does not exceed 4 Hz—the transmission bandwidth of the filters placed in the spectrum-analyser channels.

As a result of the radar observations carried out in 1961 in the Soviet Union, the signals reflected from Venus were detected and their parameters measured, making it possible to estimate the effective area of Venus for back scattering, the rotational period of Venus, the distance of Venus from the Earth and Venus' radial velocity, as well as to refine previous estimates of the astronomical unit. The astronomical unit was found to be 149,599,300 km (assuming the speed of light is 299,792.5 km/s), with a root-mean-square uncertainty of 650 km. The rotational period of Venus determined from the broadening of the reflected signal was found to exceed 100 days, if the reflective properties of Venus were similar to those of the Moon.

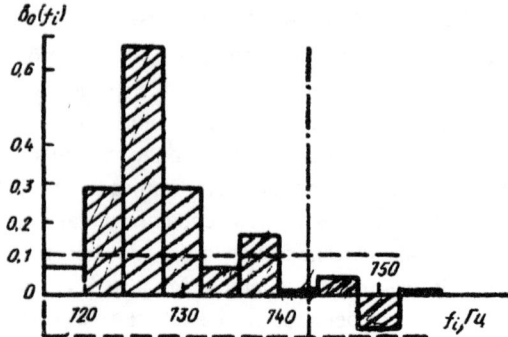

Fig. 9.3 Mean spectrum of the signal reflected from Venus, obtained on April 18, 1961: f_i are the frequencies of the spectral components of the signal at the receiver output in Hz; $b_o(f_i)$ is the ratio of the mean power of the signal in a given frequency channel of the analyser to the spectral power density of the noise; the transmission bandwidth of the filters in the analyser channels was 4 Hz, and the repetition period of the pulses and pauses in the signal was 0.256 s

In 1961, successful observations of Venus were also conducted by several radar installations in the USA and England at wavelengths from 12.5 to 74 cm, and gave results close to our own.

Radar Observations of Venus in 1962 In 1962, the energy potential of the radar installation was increased sixfold by employing a quantum paramagnetic amplifier (QPA) at the receiver input, and also by increasing the power of the transmitter. The QPA was developed by A. V. Frantsesson under the supervision of M. E. Zhabotinskii. At that time, the energy potential of the DSCC radar installation was higher than those of foreign installations.

It became possible to carry out extended and more precise measurements of the distance between the Earth and Venus, and to determine important physical characteristics of the planet. A periodic linear frequency modulator was used to enhance the accuracy of the distance measurements. Academician Kotel'nikov proposed a method providing strict linear variation of the instantaneous frequency-modulation oscillations, which B. I. Kuznetsov and Lishin used as a basis for the construction of a special generator. As was shown by Kotel'nikov in 1962, the periodic linear frequency modulation made it possible to construct a two-dimensional radar map of the surface of the planet.

The root-mean-square uncertainty for the unified measurements of the distance to Venus in 1962 did not exceed 15 km, while it had been 2700–4000 km in 1961. The 1962 observations indicated the value of the astronomical unit to be $149{,}597{,}900 \pm 250$ km. It was also found that Venus was located 270 km ahead of its ephemerides position in its orbit relative to the Earth, and that the radius of Venus was 80 km smaller than the value of 6100 km that had been accepted earlier. Similar refinements to the ephemerides were found by American researchers.

Using linear frequency modulation, the energy distribution of the reflected waves was first obtained using the linear frequency modulation in 1962, from the high-resolution delay information, which made it possible to determine the nature of

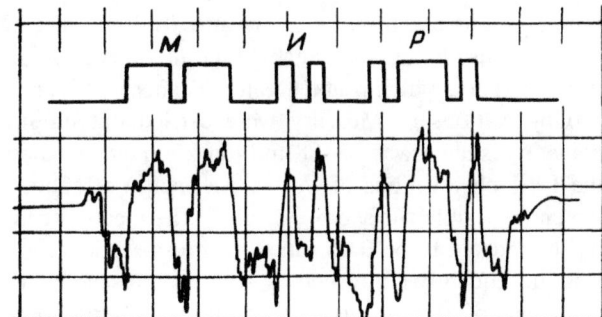

Fig. 9.4 The Russian word "MIR" ("Peace," or "World") transmitted by Morse code via reflection off Venus on November 19, 1962. The duration of a dot is 10 s and of a dash is 30 s

the reflective surface of Venus. A reflective patch was detected at the centre of the visible disc of the planet, where the waves are incident perpendicular to the surface. This phenomenon was expressed even more clearly for Venus than for the Moon.

The ratio of the effective area for the back scattering off Venus to the area of the planet's perpendicular cross section varied from 0.12 to 0.18, which is, on average, about twice that for radar observations of the Moon at the same wavelength. These data suggest that the relative dielectric permeability of the surface layer of Venus is four to six, which corresponds to the value for terrestrial rocky material in a dry state.

These investigations established that the reflection of decimetre radio waves is due to the presence of a hard surface on Venus. The atmosphere of Venus proved to be transparent to these waves, enabling the researchers involved to obtain the first evidence about the surface of the planet.

One of the very important results of the radar observations in 1962 was determining the period and direction of rotation of Venus, which were derived from day-to-day variations in the width of the spectrum of the reflected signal, which is proportional to the angular speed of the apparent rotation of the planet relative to the radar antenna.[2] It was found that the spectral width of the reflected signal is minimum at the inferior conjunction, which demonstrates that Venus' rotation is retrograde (opposite to the planet's motion around the Sun). The variation in the width of the spectrum corresponds to a rotational period of about 300 Earth days. Based on observations made in the USA in 1962, the rotational period of Venus was estimated to be about 250 days (also for retrograde rotation).

The fairly high signal-to-noise ratio that was obtained when Venus was located relatively near the Earth made it possible to use Venus as a passive reflector for telegraphic transmissions. A message transmitted on November 19, 1962 using this telegraphic method, which traversed a path of 82 million kilometres, is shown in Fig. 9.4.

---

[2]This apparent rotation was comprised of two parts: the rotation of the planet itself relative to the fixed stars and the translational motion relative to the radar antenna. The first component (which we wish to determine) is constant, while the second (variable) component can be calculated ahead of time.

Observations of Mercury, Mars and Jupiter The enhanced energy potential of the radar installation also enabled the first detection, in 1962, of a reflected signal from Mercury, which is at a greater distance and is smaller in size than Venus.

Observations of Mercury were carried out on June 10–15, 1962, when the planet's distance was 83–88 million kilometres—about twice the distance during the 1961 observations of Venus. An energy of about one Watt was incident onto the entire visible surface of Mercury. The mean effective area for the back scattering turned out to be 0.06 times the cross-sectional area of the planet. This result was subsequently confirmed by American researchers, who carried out radar measurements of Mercury using the Goldstone Deep Space Network antenna in 1963. The independent measurement of the astronomical unit obtained from the Mercury radar experiments confirmed the value that had been obtained from the observations of Venus.

The first experiments attempting to detected reflected radar signals from Mars and Jupiter took place in 1963.

Mars and the Earth were positioned at opposite ends of their orbits in 1963, and the distance to Mars exceed 100 million kilometres, providing unfavourable conditions for radar measurements. With the available sensitivity, it was expected that detection of the reflected signal would be possible only if there were present on the Martian surface fairly flat, horizontal regions that would not cause appreciable broadening of the reflected signal, despite the rapid rotation of the planet. Observations of Mars were carried out on four nights on February 6–10, 1963.

Our observations carried out later using a radar system with a higher energy potential demonstrated that the frequency filters used in February 1963 for the analysis of the reflected signal were not optimal, casting doubt on the conclusions that had been drawn about the reflective properties of the surface.

In our 39-cm observations of Jupiter in 1963, the estimated energy of the received reflected signal for an integration time of 22 hours was 1–1.5 times the root-mean-square uncertainty due to noise fluctuations. The low level of the reflected signal suggested that decimetre radio waves are nearly completely absorbed in the deep atmosphere of Jupiter, and are not scattered into the surrounding space.

Simultaneous with the Jupiter radar experiment in October 1963, Trunova and Rzhiga carried out measurements of the intrinsic radio emission of Jupiter, which had not yet been observed at 30–50 cm at that time. The equivalent temperature of Jupiter at 39 cm reduced to the visible disk of the planet was $12,000 \pm 2,000$ K. This result confirmed that the sharp growth in Jupiter's emission detected at shorter wavelengths continued at decimetre wavelengths.

The first radar studies of the planets were very well known, not only in the Soviet Union, but also abroad. In 1964, V. A. Kotel'nikov, V. M. Dubrovin, M. D. Kislik, V. P. Minashin, V. A. Morozov, G. M. Petrov, O. N. Rzhiga and A. M. Shakhovskoi were named laureates of the Lenin Prize for their radar investigations of Venus, Mars and Jupiter.

Computer Reduction of the Observational Results On June 12, 1964, with this goal, several sets of data for reflected signals from Venus were recorded on magnetic tape in digital form. These data were processed using a BESM-2M computer, which calculated the coefficients of the Fourier series and estimated the spectral density of the signal power averaged over the observations. The computer program was developed by Yu. N. Aleksandrov and V. A. Zyatitskii. The apparatus used to digitally code and record the signals onto magnetic tape, and then to transfer the data to the computer, was designed by A. I. Smurygov and N. M. Bondarenko. Features due to regions of the surface of Venus with higher reflectivity than the surrounding areas were detected in the spectrum of the reflected signal.

In essence, this was one of the first applications of the method of aperture synthesis in radar, making it possible to obtain high special resolution. The detection of high reflectivity on the surface of Venus opened the possibility of radar mapping of the surface, and also refining the rotational parameters of the planet based on the angular distribution of surface features, as can be done for planets whose surfaces are accessible to optical observations.

Joint Observations of Venus with Jodrell Bank Observatory In Summer 1963, at the invitation of the USSR Academy of Sciences, Bernard Lovell, the Director of the Jodrell Bank Observatory of Manchester University, visited the Deep-Space Communications Centre. Lovell was very impressed by the technical outfitting of the DSCC. Soon after this visit, Lovell approached Academician Kotel'nikov with a proposal to carry out joint radar observations of Venus.

The use of the Jodrell Bank 76-m-diameter MkI radio telescope as the receiver tripled the sensitivity of the radar measurements. In addition, continuous signal reception over an extended period of time became possible, since the transmitter was below the horizon, and so did not give rise to any interference.

The DSCC transmitter operated for several hours a day at a constant frequency. A reflected signal from the Moon was detected at Jodrell Bank on December 21, 1965. The first successful detection of a radar signal from Venus was on January 9, 1966. The signal frequency turned out to be below the nominal value of 30 Hz, and the signal was detected only after a frequency search was carried out at Jodrell Bank. Joint observations of Venus were carried out until March 1966.

The processing of the resulting recordings at IRE was carried out on a BESM-2M computer using programs developed by Aleksandrov and Zhurkina.

The higher signal-to-noise ratio and lowering of the fluctuation noise using long integration times enabled more reliable determination of the positions of spectral features. Using the measured radial velocities, V. K. Golovkov was able to fix the centres of regions with enhanced reflectivity on the surface of Venus.

Very good agreement was found for spectra obtained at the inferior conjunctions of 1964 and 1966, indicating that Venus turned roughly the same face towards the Earth during these different lower conjunctions. After identifying spectral features in these observations, Golovkov derived the refined estimate of the rotational period of Venus, 243.9 ± 0.4 days.

Thus, it was established that Venus' rotational period was close to the value 243.16 days for which the planet would turn exactly the same hemisphere toward the Earth at all inferior conjunctions (a situation called synodic resonance). In the interval between conjunctions, which repeat, on average, every 583.92 days, an observer on Earth would see four full rotations of Venus if its atmosphere were optically transparent. The mean duration of a solar day on Venus is 116.8 Earth days. Attempts to determine the period and orientation of Venus' axis from optical observations were made in the past, but only the radar technique enabled the acquisition of reliable data on the rotation of Venus, as well as the direction of the planet's rotational angular momentum.

Since the rotation of Venus was very important from the point of view of the evolution of the solar system, work on refining the rotational period was continued. Reduction of observations carried out in the Soviet Union, including measurements made in 1977, yielded the value 243.04 ± 0.03 days for the rotational period of Venus.

Observations of Venus and Mars During the Flight of Unmanned Spacecraft
In view of the large uncertainties in the ephemerides calculated using classical theories, beginning in 1969, the DSCC radar installation was used to conduct regular measurements of the distances and radial velocities of planets, with the aim of predicting their positions during the final approaches of unmanned interplanetary spacecraft, as well as refining the theory of planetary motions. The improved program for the computation of the ephemerides for the time delay of the reflected signals and Doppler corrections to the transmitter carrier frequency were developed by Yu. K. Naumkin under the supervision of E. L. Akim and V. T. Geraskin.

By the beginning of the 1969 radar observations, Aleksandrov, with the help of R. A. Andreev, had developed equipment for digitally recording the reflected signals and a program for their reduction on the M-220 computer of the DSCC. A fast-Fourier transform algorithm was used to reduce the processing time. This program was able to process a 5-minute recording in 30 seconds. In 1965, Petrov developed a specialised, multi-channel digital device designed to observe the spectrum of the reflected signal during reception and obtain estimates of the spectral parameters.

The digital-recording complex was applied to the analysis of radio signals emitted by the *Venera-7* spacecraft during its landing on the surface of Venus in December 1970. It was established that the spacecraft had successfully landed on Venus, and data transmitted from the surface were received at the Earth, although the power of the received signal after landing fell by several orders of magnitude.

By the time of Mars observations take in 1971, the energetic potential of the radar equipment had been increased by an order of magnitude since the first attempts to detect a signal reflected from Mars in 1963, due to an increase in the transmitter power, improvement of the antenna, which was now able to receive reflected waves with the opposite circular polarisation, and replacing the radiometer receiver with a direct comparison of the signal spectrum with a noise spectrum. The power of the reflected signal was enhanced by another factor of ten due to the fact that Mars was nearly twice as close to the Earth in the grand opposition of 1971.

In spite of these measures, the reflected signal was so weak that its spectral density at maximum was only 0.5–2% of the spectral density of the receiver noise. In order to obtain a reliable detection of the signal, it was necessary to integrate the signal energy over the entire interval when the planet was visible. In connection with this, a method was developed to take into account inhomogeneities in the frequency characteristics of the receiver tract, which could easily be mistaken for the spectrum of the reflected signal in the case of long integrations.

Observations of Mars were carried out from June 14 to September 19, 1971 in groups of three observing days each. On the first day, the continuum signal at a constant frequency was studied and the reflected signal detected. On the second day, a signal with a linear frequency modulation with deviations of 4 kHz was used, enabling measurement of the distance with an accuracy of 20 km. On the third day, the frequency deviations were increased to 32 kHz, corrections determined from the data obtained on the previous day were applied to the ephemeris values for the time delays and new measurements aimed at determining the distance to within 5 km were carried out.

J. Clemens developed a theory for the motion of Mars based on the results of these measurements. It was found that Mars was located closer to the ephemeris positions before than after the opposition. In the observed interval, the deviation of the distance from the ephemeris values varied linearly with a rate of about 2 km/day.

In the middle of 1971, at the request N. A. Savich, who was supervising studies of the Martian ionosphere, help was given with the recording and processing of radio signals at two coherent frequencies transmitted through the ionosphere. Dubrovin and Aleksandrov provided an apparatus for magnetic recording and digital processing of the signals, making it possible to carry out such observations when the *Mars*-2 spacecraft was setting behind the planet. In 1974, during the flights of the *Mars*-4 and *Mars*-6 spacecraft, this equipment was used in experiments leading to the discovery of a night-time ionosphere on Mars, and yielded measurements of the pressure and temperature on the planet's surface. It was also used to determine the characteristics of the Venusian ionosphere during the *Venera-9* and *Venera-10* flights in 1975–1976. Petrov and A. L. Zaitsev developed a digital Doppler synthesiser with quadratic frequency variation for this same experiment.

Observations of Planets Using the New Radar Installation of the DSCC Possibilities for planetary studies using radar measurements expanded considerably in connection with the construction of a new fully steerable 70-m-diameter antenna under the supervision of V. A. Grishmanovskii. The increase in the effective area of the antenna, lowering of the noise temperature of the antenna-feeder tract and receiver and increase in the power of the transmitter, which were supervised by I. E. Mach, made it possible to increase the energy potential of the radar system at 39 cm by a factor of 50. An energy of about 250 W now fell onto the visible surface of Venus at lower conjunction. The maximum distance feasible for radar measurements was increased by more than two and a half times. This meant that it was now possible to observe Venus virtually during its entire orbit.

Petrov, Zaitsev and A. F. Khasyanov radically modernised the radar apparatus, including the digital synthesiser for the linear frequency modulation and the programming-time device—the radar itself.

The improved capabilities of the radar installation enabled a series of observations of Venus, Mercury and Mars from February through April 1980, during the

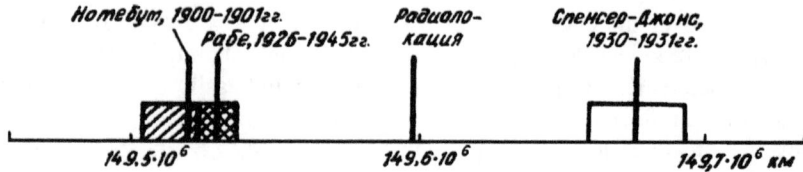

Fig. 9.5 Results of determining the astronomical unit from the parallax shift of the asteroid Eros and from radar observations. The errors in the radar values cannot be seen in the figure because they are smaller than the thickness of the line. The *horizontal axis* shows distances in km. The four measurements, *from left to right*, are those by Nomebum (1900–1901), Rabe (1926–1945), radar, and Swenson and Jones (1930–1931)

Venus observations, the uncertainty in the distance was 300–500 m at an overall distance of up to 140 million kilometres; this was made possible by taking into account the instrumental function of the radar and the beam for the back scattered signal from the planetary surface when analysing the spectrum of the reflected signal, a technique that was developed by Khasyanov. The achieved accuracy enabled continuation of studies of the surface profile of Venus that had been begun earlier, and led to the discovery of two extensive mountainous regions with altitudes of 2.5–4 km.

The radar observations of Mars and Mercury in 1980 were markedly improved from those made earlier in 1971 and 1962. With the new antenna and transmitter, it became possible to measure the distances to the surfaces of these planets with a root-mean-square uncertainty of 0.6–1.5 km in 5 to 15 minutes at an overall distance of 100–135 million kilometres. These measurements enabled refinement of the surface profile of Mars, in particular, in the region of the extinct volcano Olympus Mons—the largest known mountain in the solar system, with an altitude of 27 km.

Joint Refinement of Astronomical Constants. Construction of a Unified Relativistic Theory for the Motions of the Inner Planets One of the main results of the first radar observations of Venus that was extremely valuable for manned space flight was the refinement of the astronomical unit—the distance from the Sun to the Earth. Observations of the parallactic shift of the small asteroid Eros, which had previously been the main means of determining the astronomical unit, yielded various estimates that differed by several hundreds of thousands of kilometres, which was large enough to threaten unavoidable misses of space stations (Fig. 9.5). The value of the astronomical unit obtained from radar observations in 1961 and 1962 in the Soviet Union, England and the USA were in good mutual agreement. The XII General Assembly of the International Astronomical Union in 1964 recommended the radar value of the astronomical unit for use in astronomical almanacs.

At the same time, the radar observations showed that, even after the associated corrections to the astronomical unit, there remained appreciable systematic discrepancies between the actual and ephemerides positions of Venus relative to the Earth, reaching several hundreds of kilometres. Similar discrepancies were observed in the position of Mars.

Studies aimed simultaneously at refining several astronomical constants using radar observations were carried out in the Soviet Union and the USA. Petrov used radar data for Venus obtained in the Soviet Union in 1962 and 1964 to derive a more accurate value for the radius of the surface of Venus, 6046 ± 15 km.

Based on the accumulated observational material, work was carried out in the Soviet Union to refine the orbits of the Earth, Venus, Mars and Mercury, and to develop a theory for the motions of these planets that would provide predictions of their relative distances that were much more accurate than those given by the analytical theory of S. Newcomb created in the late 1800s.

The observations of Venus, Mars and Mercury in 1980 appreciably augmented the results of earlier radar observations of these planets, especially for Mercury and Mars, for which the last available data were published in 1965 and 1971. This provided a basis for constructing a unified theory of the motions of the inner planets of the solar system, in other words, for finding a simultaneous solution for the orbital elements for the motions of Mercury, Venus, the Earth and Mars in a time interval of about 20 years.

The main defining characteristic of the new computation method was the use of relativistic differential equations to describe the heliocentric motions of the planets.

The root-mean-square deviations of the measured distances beginning in 1967–1970 were 0.9 km for Venus, 2 km for Mercury and 2.5 km for Mars. Taking into account the surface profiles of the planets decreased the deviations for Venus to 0.5 km and for Mars to 1 km. At the same time, the root-mean-square deviations of the results of optical measurements in this time interval are virtually constant and about 100 times larger when translated into linear distances.

The achieved agreement between the experimental and computed data provides experimental verification using direct astronomical methods of the general theory of relativity, which encompasses all possible relativistic effects in the motions of the planets and the propagation of electromagnetic waves. Radar observations of Venus carried out in the Soviet Union from December 1981 until February 1982 confirmed the high accuracy of the theory. The discrepancies between the predicted and measured distances did not exceed 1.2 km.

In 1982, Yu. N. Aleksandrov, V. K. Abalakin, V. A. Brumberg, M. D. Kislik, Yu. F. Kolyuka, G. A. Krasinskii, G. M. Petrov, Gr. M. Petrov, V. A. Stepan'yants, K. G. Sukhanov, V. F. Tikhonov and A. M. Shakhovskoi were awarded a State Prize of the USSR for establishing a unified relativistic theory of the motions of the inner planets of the solar system.

Investigations of the Integrated Characteristics of the Scattering of Radio Waves from Planetary Surfaces Measurement of the polarisation of the reflected radiation and the distribution of its intensity over the disk of the planet demonstrated that, in contrast to optical radiation, radio waves reflected from the Moon and planets display primarily a mirror character.

At the same time, part of the radiation is reflected diffusely. In 1968, Aleksandrov and Rzhiga analysed this diffuse component, and showed using Venus as an example that the relative dielectric permeability of the planetary surface calculated using the mirror-reflection coefficient was 20% lower than was obtained by assuming that all the 39-cm radiation was mirror-reflected. It was shown that the relative dielectric permeabilities of various regions of the Venusian surface derived from the 39-cm observations were in the range 2.7–6.6, and the corresponding densities in the range 1.3–3 g/cm^3. These data were subsequently confirmed by surface-density measurements made from interplanetary spacecraft.

A complex analysis of the physical properties of the surface layer of Mars based on radar, radio astronomy and infrared observations was carried out by Rzhiga in 1967.

In 1968, Aleksandrov, B. I. Kuznetsov and Rzhiga showed that the characteristics of the mirror-reflected radiation from Venus were consistent with a model with a wavy, uneven surface, with an exponential spatial altitude auto-correlation function; the diffuse component corresponded to Lambert scattering. The most probable inclination of the Venusian surface implied by the 39-cm observations was, on average, 2.6°. The surface of Venus is smoother than the surface of the Moon, but rougher than the surface of Mars.

Studies of the Absorption and Refraction of Radio Waves in the Atmosphere of Venus Methods for analysing the reflection characteristics of planets were applied by Aleksandrov and Rzhiga in 1968 to investigate limb-darkening due to the absorption of the reflected radio radiation in the atmosphere of Venus. It was shown that this limb-darkening was consistent with the atmospheric absorption that was implied by a decrease in the effective area of Venus detected in radar observations at centimetre wavelengths carried out by the Lincoln Laboratory of the Massachusetts Institute of Technology. This fact was very important for elucidating the origin of the observed decrease in the reflectivity of the planet.

Also in 1968, Rzhiga showed that the back-scattering measurements were in better agreement with the hypothesis of non-resonance absorption of the radio waves in gaseous components in the atmosphere of Venus. In 1970, he constructed a model adiabatic carbon-dioxide atmosphere, which predicted that the atmosphere of Venus should give rise to an additional delay of about 2 milliseconds in radar signals.

The measurements of the *Venera-4* spacecraft ceased at an altitude above the planetary surface where the pressure was about two million Pascals (MPa). Assuming that the absorption indicated by the radar data was due entirely to carbon-dioxide gas and water vapour, and extrapolating the atmosphere's parameters along the relation for an adiabatic gas, Rzhiga found that the mean surface pressure on Venus was 11.1 ± 3.0 MPa, while the mean surface temperature was 740 ± 35 K. These results were taken into account in the construction of the *Venera-7* spacecraft. The actual pressure and temperature measured by the *Venera-7* and *Venera-8* spacecraft were 9.1–9.6 MPa and 740–750 K, confirming the earlier estimates.

Developing the original ideas of L. I. Mandel'shtam and N. D. Papaleksi, and continuing from the first lunar radar experiments of the Gorkii Radio Physics Research Institute, the Institute of Radio Technology and Electronics of the USSR Academy of Sciences, in collaboration with a number of other institutions that participated in the development and outfitting of the radar system and in observations with the system, in particular with colleagues at the DSCC in the Crimea, established a new direction in astronomy in the Soviet Union—radar planetary astronomy.

Beginning in April 1961, with the first detection of reflected signals from Venus, a number of fundamental results were obtained. The value of the astronomical unit was substantially refined. The rotational period and direction of rotation of Venus were determined. The distribution of the reflectivity across the disk of Venus was obtained, as well as the degree of polarisation of the reflected signal, making it possible to establish the nature of the Venusian surface. Reflected signals from Mercury were detected for the first time. A unified relativistic theory for the motions of the inner planets that enabled computation of their mutual positions with an accuracy 100 times better than provided by classical theories was obtained.

9.2 Studies of the Earth and Planets Using Radio Physical Methods

Studies of Thermal Radio Emission of the Earth and planets were begun at IRE in 1960. The main initiator and supervisor of this work was Professor A. E. Basharinov (1920–1978).

Interest in studies of the thermal emission of the atmosphere using radio astronomy methods at IRE was associated with the problem of establishing the relationship between the noise emitted by the atmosphere at millimetre wavelengths and meteorological parameters. Objects of study included not only the cloudless atmosphere, but also clouds, precipitation, and turbulent inhomogeneities. Further objects added to this list include ice, the sea surface, ground vegetation, the soil, etc. These investigations represented a development of studies on diagnostics for low-temperature plasma formations, begun earlier at the suggestion of Basharinov and carried out by V. M. Polyakov and his coworkers.

Radiometer studies of natural objects required the development of high-sensitivity microwave radiometers operating at a wide range of millimetre and centimetre wavelengths. At the end of the 1950s, the Specialised Construction Bureau (SCB) was organised at IRE. One direction pursued by this organisation was the development and manufacture of microwave radiometers. A large role was played in this work by V. S. Ablyazov. In the 1960s and 1970s, radiometers for discrete spectral intervals from 3 cm to submillimetre wavelengths (0.8, 1.35, 2.25 cm and others) were developed at the SCB, as well as universal low-frequency modulational-radiometer blocks and other such equipment, which were manufactured in modest quantities. For a long time, the IRE Specialised Construction Bureau was one of only a few organisations specialising in the development of radio astronomy equipment. Most radio telescopes in the Soviet Union have receivers developed at the SCB. Microwave radiometers intended for radio astronomy studies of the atmosphere and surface of the Earth to be carried out on board airborne laboratories and spacecraft have recently been developed under the supervision of Ablyazov. Flying IRE laboratories carried by IL-18, IL-14, and AN-2 aircraft are outfitted with these radiometers. A polarisation-sensitive 2.2-cm microwave radiometer was installed on board the *Interkosmos-20* and *Interkosmos-21* spacecraft, and operated reliably for a long time (Ablyazov, V. P. Bydantsev).

Studies of the thermal radio emission of an atmosphere containing clouds, precipitation, and turbulent inhomogeneities require measurements of radio brightness temperatures and the total absorption in the atmosphere simultaneously at several wavelengths in the millimetre and centimetre ranges. Such experiments were first conducted by IRE staff using the 22-m telescope of the Lebedev Physical Institute, with active help and support from A. E. Salomonovich and R. L. Sorochenko. Experimental studies from the surface of the Earth revealed absorption features in the microwave spectra and the dependence of the brightness temperature of the cloudy atmosphere on temperature, the water content of the clouds, the water droplet size, the intensity of rain, and so forth. In addition, radio brightness contrasts for different types of clouds and rain were measured, and data on the spatial and temporal dependences of the intensity of fluctuations of the radio emission of the atmosphere were obtained. These results were used to construct theoretical models for the radio emission of a cloudy atmosphere and to develop methods for determining meteorological parameters from measurements of the spectral and polarisation characteristics of atmospheric emission (Basharinov, B. G. Kutuza).

The possibility of remote sounding of the atmosphere from spacecraft called forth new interest in studies of the radio emission of the atmosphere in the middle of the 1960s. In this connection, an important role was played by the work of Basharinov, S. T. Egorov, M. A. Kolosov, and Kutuza, which laid out the principles of microwave radiometer methods for deriving atmospheric parameters from flying equipment.

In the middle of the 1960s, investigations of the thermal radio emission of the sea surface and continents were carried out at IRE, leading to the development of microwave radiometer methods for deriving the parameters of the sea surface (Basharinov, A. M. Shutko).

Ground-based experimental investigations of the radio emission of areas covered by ice at centimetre and decimetre wavelengths displayed the dependence of the brightness temperature on the contrast, thickness and temperature of sea ice. The influence of floating ice on the radiative efficiency of the sea surface was estimated, and means of determining the parameters of sea and shelf icebergs based on studies of their thermal radio emission were studied (Basharinov, A. A. Kurskaya).

The first attempt to realise microwave radiometer methods on board a spacecraft was made with the goal of investigating the atmosphere and surface of Venus, about which very little was known at the beginning of the 1960s. Interest then turned to studies of the cosmic radio emission of the Earth as a planet. In 1968, the *Kosmos-243* spacecraft was launched, with a four-channel radio telescope directed toward the Earth on board. Measurements were carried out at wavelengths of 8.5, 3.4, 1.35 and 0.8 cm, with the spatial resolution provided by the antenna corresponding to 20 km on the surface of the Earth. This experiment was prepared and outfitted by groups of the IRE, the Institute of Atmospheric Physics of the USSR Academy of Sciences and industrial organisations, under the supervision of Basharinov, A. S.

Gurvich and Egorov. The space experiment on board the *Kosmos-243* spacecraft was four years ahead of foreign investigations in this area, and showed the effectiveness of microwave radiometer methods for remote sounding of the atmosphere, dry land and the world ocean. As a result of this experiment, the water content of clouds in the form of vapour and liquid droplets was determined, the total humidity of the atmosphere estimated and zones of precipitation traced on a global scale. Measurements above regions of sea yielded data on the latitude distribution of the temperature of the surface of the world ocean, and enabled storms and floating ice to be distinguished, independent of cloud cover. Observations of the continental landmasses made it possible to distinguish zones of humidity on the Earth's surface, and to detect the anomalously low brightness temperatures of the shelf and continental ice masses of the Antarctic.

In recent years, microwave radiometer methods for studying the atmosphere and world ocean have been applied in experiments on the oceanographic spacecraft *Kosmos-1076* and *Kosmos-1151*. A distinguishing feature of these experiments was the joint use of spectral and polarisation measurements of the Earth's radio emission and the enhanced sensitivity of the receivers used, which appreciably expanded the informational potential of the equipment. These microwave radiometers were developed under the supervision of Egorov. The scientific programme of these measurements was compiled at IRE, jointly with the Sea Hydrophysical Institute of the Ukrainian SSR Academy of Sciences and the Dniepropetrovskii State University (B. A. Nelepo, N. A. Armand, B. E. Khmyrov, Yu. V. Terekhin, B. G. Kutuza, E. I. Bushuev).

Studies of the radio emission of the planets were begun at IRE at the beginning of the 1960s, under the supervision of Basharinov and Kolosov. G. M. Strelkov proposed a model for the atmosphere and surface of Venus, and on its basis carried out computations of the spectrum of the emergent radiation as a function of temperature, pressure and other physical conditions at the planetary surface. Basharinov and Kutuza attempted to explain a feature in the radio spectrum of Venus as due to the presence in the planet's cloudy layer of supercooled water droplets, whose microwave absorption spectrum had been studied in a preliminary way under the conditions of the Earth's atmosphere.

Possibilities for studying the radio emission of the planets are expanding substantially with the installation of radio telescopes on spacecraft. Experiments designed for studies of Mars at 3.4 cm were installed on board the *Mars-3* and *Mars-5* Soviet interplanetary space stations, in which IRE took part, together with the Space Research Institute, Lebedev Physical Institute and the Radio Physical Institute in Gorkii. During a close flyby of the planet, the spatial resolution achieved by the radio telescope was 100–400 km. Data on the variability of the dielectric permeability of the Martian surface layer and variations in the temperature distribution over the surface were obtained (see Essays 1 and 2 in this collection). IRE participants in these studies included Basharinov, Kolosov and Shutko.

The experiments conducted on board airborne laboratories showed the potential of applying microwave radiometers to a number of national economic tasks, such as determining the ice conditions in the Arctic during ship passages, detecting active regions in forest fires under complex meteorological conditions, estimating expected harvests, determinating the water content in the atmosphere and soil, etc. The most complete of these is the development of a method for routinely determining the humidity of ground soils (Basharinov, Armand, Shutko). A service for the routine microwave radiometer monitoring of soil humidity on board AN-2 aircraft was established in Moldavia, and maps of the hydrological conditions in territories adjacent to the Karakum Canal were made in Turkmenia. Experimental verification of this approach to routine determination of ground-soil humidity in fields in various regions of the Soviet Union have demonstrated the high efficiency of this method.

Investigations Carried out Using Radio Waves Emitted by Spacecraft[3] The space age brought intense studies of the Moon and other planets of the solar system using various radio-physical methods. The idea behind such methods is to measure the characteristics of radio waves emitted by a transmitter on a spacecraft after their passage through the troposphere and ionosphere of the Moon or a planet (akin to radiography), or after their reflection from the surfaces of these bodies. Such studies began at IRE in 1962, and were carried out until 1982 under the supervision of Professor M. A. Kolosov (1912–1982). The methods most applied in practice were bistatic radar, proposed by O. I. Yakovlev (1965), the adaptation of the radiographic method initially developed by Yakovlev (1964) to make it suitable for studies of the surface of Venus by A. G. Pavel'ev (1974) and dispersion interferometry, proposed by N. A. Savich (1963).

The first experimental studies using the *Luna-11* and *Luna-12* spacecraft were conducted in 1966 using the bistatic (or bipositional) radar technique, based on analysing signals transmitted from a spacecraft after they have been scattered by a surface, with the observations carried out on the Earth. Dependences of the reflection coefficients for the scattering of metre-wavelength radio waves on the angle of incidence were obtained, making it possible to determine the mean dielectric permeability ($\varepsilon = 2.8$) and density ($\rho = 1.4$ g/cm^3) of the lunar rocks. These studies were continued at metre and decimetre wavelengths during the period when the *Luna-14* and *Luna-19* spacecraft were operational. The relationship between the shape of the energy spectra of the reflected signals and characteristic features on the lunar surface was also established.

Further, this method was successfully applied to study the surface of Venus as well. Experiments were conducted in 1975 using the *Venera-9* and *Venera-10* spacecraft. Processing of the data obtained provided new information about a number of important characteristics, such as the dielectric permeability of the Venusian soil, the density of surface rocks, the slopes of small-scale inhomogeneities on the surface, and the deviations in altitude from a spherical surface associated with large-scale components of the surface profile. Radar maps of the surface of the planet were also produced.

The first radiographic experiment involving the transmission of radio signals through the atmosphere of Venus was carried out in 1967, during the descent of

---

[3] A. I. Efimov took part in the preparation of material for this section.

the *Venera-4* interplanetary station to the planetary surface. This provided information about the conditions for the passage of the radio waves through the Venusian troposphere, as well as data on the turbulency of this medium. Similar experiments conducted using the descending *Venera-5* and *Venera-8* spacecraft made it possible to more accurately determine the characteristics of inhomogeneities in the Venusian troposphere. A large volume of data on the atmosphere of Venus was obtained during the operation of the *Venera-9* and *Venera-10* spacecraft in 1975–1976. In particular, altitude profiles of the temperature, pressure and particle number density were derived, and the dependences of the structural characteristics of the index of refraction on altitude in the atmosphere, the Venusian latitude and the zenith angle of the Sun were determined.

The radiographic method was also successfully applied to studies of the rarefied troposphere of Mars using the *Mars-2*, *Mars-4* and *Mars-6* spacecraft. The altitude dependence of the pressure and temperature was measured at several locations on the planet, as well as variations in these parameters with variations in the solar zenith angle.

The radiographic and dispersion-interferometry methods were used to study the characteristics of plasma envelopes surrounding the Moon and planets. The latter technique could also be considered a multi-frequency radiographic method, with all the signals analysed at the various frequencies being coherent. This approach proved to be especially effective means of studying rarefied plasma envelopes, such as night-time planetary ionospheres and the plasma surrounding the Moon. The first radiographic experiment carried out near the Moon involved decimetre and centimetre signals transmitted during the flight of the *Luna-19* spacecraft in 1972. A thin layer of plasma several tens of kilometres thick with a maximum particle density of 10^3 cm^{-3} and lying at an altitude of 10 km was discovered above the side of the Moon illuminated by the Sun. Such experiments were continued in 1974 using the *Luna-22* spacecraft. The altitude profiles of the electron density for various strengths of illumination of the lunar surface were derived.

The Martian ionosphere was probed using coherent decimetre- and centimetre-wavelength radio signals transmitted by the *Mars-2*, *Mars-4* and *Mars-6* spacecraft during 1971–1974. The most important result to come from these experiments was the detection of the night-time ionosphere of Mars. It was established that this ionosphere is strongest at a well defined altitude of 110–130 km above the planetary surface, where it has an electron density of about 5×10^3 cm^{-3}. The day-time ionosphere of Mars has an electron density of 2×10^5 cm^{-3} and is located at an altitude of about 145 km.

The dispersion-interferometer method was used to study the ionosphere of Venus during the flights of the *Venera-9* and *Venera-10* spacecraft. The altitude profile of the electron density was found above various regions of the planet. The night-time ionosphere of Venus was found to have regions of maximum electron density at two well defined altitudes.

Researchers at IRE and other organisations, including M. A. Kolosov (the project leader), N. A. Armand, N. A. Savich, O. I. Yakovlev, R. V. Batik'ko, Yu. I. Bekhterev, M. B. Vasil'ev, A. I. Efimov, L. V. Onishchenko, A. L. Pilat, B. P. Trusov and D. Ya. Shtern, were awarded a USSR State Prize in 1974 for their series of investigations of the propagation of radio waves in deep space using the *Luna*, *Venera* and *Mars* spacecraft.

Experience with radar observations of planets from the Earth was based mainly on radio mapping of Venus carried out in 1983–1984 using the *Venera-15* and *Venera-16* automated spacecraft (with Rzhiga as the scientific supervisor of this work). The radar system, which was developed at the Moscow Energy Institute under the leadership of Academician A. F. Bogomolov, enables imaging of the planet's surface with a resolution of 1–2 km and measurements of altitude profiles with an uncertainty of only 30 m. The synthesis of the radar images and surface profiles and reconstruction of photographic maps of Venus using digital methods is performed at IRE (Yu. N. Aleksandrov, A. I. Sidorenko). In the eight months following November 1983, the entire Northern hemisphere of Venus above 30° latitude has been imaged. Unique images of the surface showing mountain ridges, craters, plateaus, folds and fissures in the Venusian crust have been produced. Signs of tectonic activity on Venus have also been found. These data have enabled the creation of maps that can be used to study processes occurring on the surface of Venus, as well as the history of this planet.

Chapter 10
Radio Astronomy Studies of the Sun at the Institute of Terrestrial Magnetism, the Ionosphere and Radio-Wave Propagation of the USSR Academy of Sciences

S.T. Akinian, E.I. Mogilevskii, and V.V. Fomichev

Abstract The history of the development of and main results of studies of the radio emission of the Sun at the Institute of Terrestrial Magnetism, the Ionosphere and Radio-Wave Propagation (IZMIRAN) is briefly laid out, beginning with the earliest studies in the 1940s. The importance of regular observations of solar activity that have continued since that time at IZMIRAN is emphasised.

Two main important results were obtained during the Brazilian expedition of the USSR Academy of Sciences to observe the total solar eclipse of 1947 (see Essay 1): (1) it was found experimentally that an appreciable amount of the solar radio emission at metre wavelengths remained during the period when the visible solar disc was most fully occulted by the Moon, demonstrating that the region of radio emission was located fairly high above the photosphere; (2) the changes in the received radio signal with time during the eclipse[1] could be used to identify the locations of sources of enhanced radio emission, and to establish a correspondence between these sources and manifestations of solar activity observed at optical wavelengths.

While the first conclusion could be drawn immediately during the observations, the second required a comparison of the results of the radio observations with optical observations, including calculations of the expected eclipse times for various sunspots, faculae, filaments, protuberances and other manifestations of solar activity distributed on the solar surface on the day of the eclipse.[2] In this connection, after returning from the expedition, S. E. Khaikin made contact with colleagues at the Institute of Terrestrial Magnetism (NIIZM, now IZMIRAN—Institute of Terrestrial Magnetism, the Ionosphere and Radio-wave Propagation), where regular

---

[1] Given the first finding above, the solar eclipse observed at metre wavelengths effectively begins earlier and ends later than the optical eclipse (first and fourth contacts).

[2] Due to cloudy weather, it was not possible to carry out optical observations of the eclipse during the Brazilian expedition.

optical observations of the Sun were carried out in those years.[3] At his request, E. I. Mogilevskii compared the optical data and radio observations, revealing a correspondence between the eclipses of a site of radio emission at 1.5-m wavelength and of an active region in which eruptive protuberances and filaments had been observed. This confirmed the idea that the observed radio waves at such wavelengths were associated with the excitation of plasma, not at the photospheric level, but at the height of the inner solar corona. Although many further studies were required to elucidate the nature of the sources of the solar radio emission, this work contained the first sprouts of ideas that have not lost their importance to this day.

This first positive experience of a collaboration between the NIIZM and the Lebedev Physical Institute (FIAN), where solar radio astronomy was subsequently actively developed, provided a push in the development of solar radio observations at NIIZM as well. At the initiative of Khaikin and the Director of NIIZM Professor N. V. Pushkov, the Laboratory of Solar Radio Emission began to be established in the NIIZM Department of Solar Physics in the first half of 1948. It was necessary, first and foremost, to construct radio telescopes suitable for studies of the refraction of radio waves in the terrestrial atmosphere and ionosphere, and to begin systematic investigations of the radio emission of the Sun as a characteristic of solar activity. At that time, this work was headed by V. A. Baranul'ko, who organised the construction of the experimental platform and the acquisition of military radar systems together with the engineer P. I. Ishkov. It was possible to build solar radio telescopes based on the antennas and some of the receiver equipment from these systems.

Young radio technicians and engineers joined the laboratory in the Summer of 1948. Some of these (S. T. Akinian, M. V. Kislyakov, N. F. Skobelev) have remained at the Laboratory of Solar Radio Emission in the IZMIRAN Department of Solar Physics to this day. The first consultants and assistants in connection with constructing the radio telescopes were colleagues at FIAN—the then young scientist V. V. Vitkevich and Khaikin's PhD student B. M. Chikhachev. The first year of the laboratory (especially the Winter of 1948–1949) is extremely well remembered by those who took part in this work, because absolutely everything was begun from scratch: they used their own hands to build the first simple buildings in which they lived and worked, get the first radiometers and antennas properly working in the absence of microwave measurement equipment. That first year is remembered as a year of energetic activity by a small group of laboratory enthusiasts.

In the Spring of 1949, recordings of the solar radio emission at frequencies of 600 and 200 MHz were obtained. Using the radar antennas, which could rotate only in azimuth, it was possible to receive the solar radio emission for about an hour during the rising and setting of the Sun. The signals received directly and after reflection from the Earth's surface provided an interference pattern before and

---

[3]Recall that the Main Astronomical Observatory in Pulkovo and the Crimean Astrophysical Observatory in Simeiz were not operational in the years immediately following the end of the war. Regular observations of the Sun were obtained during World War II only in Tashkent and at NIIZM (in the Urals) before 1945, and then in subsequent years near Moscow, not far from the village Krasnaya Pakhra (now Troitsk) after the return of evacuees to this region.

10 Radio Astronomy Studies of the Sun at the Institute of Terrestrial Magnetism

Fig. 10.1 Antenna of the radio interferometer with its parallactic mount (frequency of 600 MHz, beginning of the 1950s)

Fig. 10.2 Co-phased antenna, operating at a frequency of 200 MHz (outfitted at the beginning of the 1960s)

during the rising and setting of the Sun, enabling measurement of the refraction of the radio waves in the Earth's ionosphere and atmosphere. The antenna mounts were later changed: the parabolic antenna of the telescope operating at 600 MHz was mounted on a parallactic-mount (Fig. 10.1), and the co-phased antenna operating at 200 MHz became steerable in both azimuth and altitude (Fig. 10.2). This enabled in 1951 the beginning of systematic, long-term observations, the accumulation of data on the solar radio emission and comparisons with optical observations of the Sun and with ionospheric and geomagnetic disturbances caused by solar activity.

This work stimulated the organisation in NIIZM of the first (1952–1953) Soviet photoelectric measurements of magnetic fields on the Sun,[4] which had been a subject of study for many years in solar research at NIIZM, including the nature of radio outbursts (in particular, type IV bursts). Systematic investigations of the spo-

---

[4]One consultant during the construction of the first modulation solar magnetograph at NIIZM was G. S. Gorelik, who was working in Gorkii at that time (which was then a closed city).

radic solar radio emission required appreciably expansion of the wavelength range covered by observations.

With the arrival at NIIZM of a group of graduates from the Gorkii State University (Ya. I. Khanin, A. K. Markeev, S. A. Amiantov and others), it became possible to turn attention to substantially expanding the research programme of the institute. It was necessary, taking into account the capabilities of the laboratory, to determine the research themes for the coming years, such that the observational material obtained enabled studies of the nature of solar radio outbursts associated with magnetic fields and optical phenomena occurring in solar active regions being studied in the department.

In addition to the operating radio telescopes, radio spectrographs and radio polarimeters were needed for metre wavelengths, where varied and complex sporadic radio phenomena were more richly manifest. In 1960, A. K. Markeev designed and constructed a radio spectrograph operating at 40–90 MHz. This spectrograph is systematically carrying out solar observations to this day (although it has been reworked and improved a number of times). Together with Markeev, new staff at the laboratory (graduates of Leningrad and Moscow State Universities G. P. Chernov and O. S. Korolev, A. N. Karachun and others) developed a radio polarimeter operating at 74 and 3000 MHz and introduced it into regular use. In this way, a complex of radio telescopes (operating primarily at metre wavelengths) and instruments was created, whose many years of use provided unique observational material on the complex and varied structure of solar radio outbursts.

With the arrival at the laboratory in 1963 of the Gorkii State University educated V. V. Fomichev and I. M. Chertok, research on physical analyses of solar radio outbursts was significantly increased. In particular, as a result of studies of the polarisation characteristics of type III bursts, it was concluded that the magnetic-field strengths above active regions decreased with distance R above the photosphere level more slowly than R^{-3}. This enabled estimation of the parameters of shock waves and the plasma density and magnetic-field strength in the corona in the framework of models in which type II outbursts and chains of type I bursts were generated by collisionless shocks.

In 1970, the experienced radio astronomer B. M. Chikhachev, who had come to work at IZMIRAN from FIAN, became the Director of the Laboratory. However, major plans for constructing new radio telescopes and establishing a radio astronomy station in the Crimea that were matured by Chikhachev were not destined to be realised. His sudden death in 1971 prevented the introduction of original, high-sensitivity methods for radio astronomy studies of the ionosphere proposed by him into practice.

Over the last ten years, the main efforts of the laboratory (whose Director is V. V. Fomichev) have been concentrated in investigations of solar radio phenomena at metre wavelengths, with the accent placed on studies of the fine structure of these events, with the aim of elucidating the mechanisms generating the bursts and obtaining information about the physical parameters of the plasma in the solar corona. The fine structure of the dynamical spectra of type IV continuum bursts (filaments, zebra structure, pulsations) was studied in detail, the fine structure of type V bursts was studied and described for the first time, new properties of type II bursts were discovered, the properties of fluctuations and quasi-periodic components in noise storms were investigated at metre wavelengths, and so forth. The experimental data obtained indicated the complex structure of the solar corona and the presence of various types of turbulence in the coronal plasma, and also made it possible to estimate certain parameters of the corona.

Another research direction of the Laboratory is associated with the physics of solar–terrestrial links. In particular, methods for quantitative diagnostics of solar proton flares were developed under the supervision of Chertok, which could be used to derive the parameters of proton fluxes with energies of more than 10, 30 and 60 MeV in near-Earth space from observations of solar radio outbursts, as well as the amount of absorption of short radio waves in the near-polar ionosphere. Data on the outbursts at centimetre wavelengths were used to obtain information about the quantity of particles accelerated by an outburst, while metre-wavelength data yielded information about the conditions for the escape of the energetic particles into interplanetary space. This method takes into account the solar longitude of an outbursts, as well as the frequency spectrum of the radio emission at centimetre wavelengths (the energetic particle spectrum). Verification of the method using independent material showed that it could be used to obtain operative estimates of the expected parameters of proton fluxes at the Earth's orbit from several to tens of hours before their actual arrival.

Although small antennas are used in the IZMIRAN complex of radio telescopes and spectrographs operating at metre wavelengths, preventing the unambiguous identification of individual sites of radio emission, regular observations with high temporal and frequency resolution analysed together with optical observations can yield high quality, and often unique, data on the fine and varied structure of solar radio outbursts and accompanying phenomena in the optical. This had led, in particular, to the active participation of IZMIRAN in international research programmes targeting solar outbursts (such as the Year of the Solar Maximum 1980–1981).

Chapter 11
The Birth and Development of Radio Astronomy Studies of the Sun at the Siberian Institute of Terrestrial Magnetism, the Ionosphere and Radio-Wave Propagation

G.Y. Smol'kov

Abstract The history of the organisation of the Department of Radio Astronomy at the Siberian Institute of Terrestrial Magnetism, the Ionosphere and Radio-Wave Propagation (SibIZMIRAN) is described, together with the principles behind the construction of the Siberian Solar Radio Telescope and the results of observations of the solar radio emission at decimetre wavelengths using this telescope.

11.1 Organisation of the Department of Radio Astronomy

Observations of the radio emission of the Sun using the Irkutsk Magneto-Ionospheric Station (IMIS) began in 1956 in connection with the *International Geophysical Year*. These observations were needed to remove the gap between the observations at the Ussuriisk and European observatories, as well as for the further development of geophysical research at Irkutsk.

In spite of the difficult conditions, the Laboratory for Solar Studies (supervised by G. Ya. Smol'kov) was organised relatively rapidly at IMIS. Regular observations of the solar radio emission began in 1958 at 209 MHz using a radiometer built at the Laboratory based on an STsR-637 radar station (Smol'kov, V. I. Burkov, N. K. Osipov and others).

Directly after this, the construction of a radio telescope operating at 10 cm wavelength began, as well as meteor radar facilities at 4 and 10 m wavelength. The radio telescope was introduced into operation in Noril'sk, and was first used for solar observations (in hopes of being able to obtain round-the-clock measurements), and then, more successfully, to record the radio noise produced by polar auroral layers of the Earth. The meteor studies were later transferred to Tomsk Polytechnical Institute in connection with the fact that the research themes of the Laboratory were restricted to solar physics.

With the transfer in 1959 of IMIS to the Siberian Division of the USSR Academy of Sciences, and the organisation on its basis of the Siberian Institute of Terrestrial Magnetism, the Ionosphere and Radio-Wave Propagation (SibIZMIRAN), in 1960 came the development and realisation of a programme for the creation of a new

Fig. 11.1 Small-baseline radio interferometer (wavelength of 3.5 cm)

scientific complex at Irkutsk, which enabled modern studies of solar activity simultaneously using all ground-based methods available (Smol'kov, G. V. Kuklin, V. E. Stepanov, R. B. Teplitskaya, V. G. Banin, V. M. Grigor'ev, V. M. Skomorovskii, A. T. Altyntsev and others).

Many radio astronomy organisations (first and foremost, the Main Astronomical Observatory; Gorkii Radio Physical Institute; Lebedev Physical Institute; Institute of Terrestrial Magnetism, the Ionosphere and Radio-Wave Propagation; and the Institute of Applied Physics in Gorkii) helped us to master various methods and observing techniques and to prepare specialists. Collaborative studies are carried out with some of these organisations to this day.

Such collaborative efforts were facilitated in the first stage of the work of the new institute by the construction in Irkutsk of a radio polarimeter operating at 3.2 cm (Smol'kov, T. A. Traskov, V. P. Nefed'ev, L. E. Treskova, 1963) and a radio interferometer with a short baseline operating at 3.5 cm (V. G. Zankanov, Treskov, M. A. Khaitov, Smol'kov; Fig. 11.1). These formed the basis for studies of the polarisation and fluctuation properties of the solar radio emission (which became traditional at Irkutsk).

In recent years, the elucidation of the nature of active regions and of phenomena leading to the development of flares required studies of their structure and the dynamics of their development with high spatial resolution. It was expedient to carry out such studies at Irkutsk, since there were already well equipped optical observatories there, which were being used to conduct multi-faceted investigations of solar activity.

In this connection, SibIZMIRAN proposed in 1964 to erect a specialised solar radio telescope in Eastern Siberia, with a spatial resolution in both directions an order of magnitude higher than other solar radio telescopes in the world, making it possible to obtain radio images of the Sun in real time. This proposal was approved by the Radio-Astronomy Scientific Council of the USSR Academy of Sciences and the Presidium of the Siberian Division of the USSR Academy of Sciences.

The type and operational principles of the Siberian Solar Radio Telescope (SSRT) were chosen together with the Pulkovo (Main Astronomical Observatory) radio astronomers D. V. Korol'kov and G. B. Gel'freikh. Plans for the SSRT were confirmed in 1965. Corresponding member of the USSR Academy of Sciences A. A. Pistol'kors was appointed the scientific supervisor for the development of the technical design of the SSRT in 1967. The technical design composed by Pistol'kors, Smol'kov and Treskov was approved in 1969, and obtained high marks from the expert committee of the Siberian Division of the USSR Academy of Sciences, chaired by Academician A. L. Mints. The decision to construct the radio telescope was made in May 1972.

An operational model of the radio telescope—an eight-element radio interferometer—was assembled in 1974 at IMIS, as a means of verifying various design elements and testing certain nodes and systems of the SSRT. This facility was subsequently extended and transformed into a ten-antenna compound interferometer in 1979 (Smol'kov, N. N. Potapov, Treskov, B. B. Krissinel', V. A. Putilov and others). A second radio polarimeter working at the operational wavelength of the SSRT was also constructed in this same location in 1977 (Potapov and others).

With the beginning of the development of the SSRT project, the volume of radio-astronomy research carried out at SibIZMIRAN increased substantially. This led to the decision by the Presidium of the Siberian Division of the USSR Academy of Sciences to provide radio astronomers their own laboratory in 1965, which was reorganised into the Radio Astronomy Department in 1980 (in connection with the introduction into operation of the SSRT).

11.2 The Siberian Solar Radio Telescope

The Siberian Solar Radio Telescope is a so-called radio heliograph, intended for studies of the structure and dynamics of the development of active regions and flares in the solar atmosphere (Fig. 11.2). Strengths of the SSRT include its high resolution in both coordinates and the possibility of rapidly and frequently obtaining radio images of the solar corona facing the Earth (this is impossible at optical wavelengths due to the very high brightness of the solar disc).

Active regions in the solar corona can display very high contrasts at wavelengths of 5–6 cm, especially in polarised light. Based on this fact, and taking into account the operational wavelengths of radio heliographs that were already available or under development, the operational wavelength for the SSRT was chosen to be 5.2 cm. The field of view of the SSRT is $35 \times 35'$, which corresponds to the radio diameter of the Sun at this wavelength.

Fig. 11.2 Siberian Solar Radio Telescope (SSRT) in the Badara thicket (East–West line, comprised of 128 antennas with diameters of 2.5 m)

The high two-dimensional spatial resolution is achieved thanks to the large synthesised aperture of the antenna system—a 256-element radio interferometer forming a single large antenna with a diameter of 622 m. This radio interferometer consists of two mutually perpendicular rows of equidistant parabolic surfaces (128 in each row) oriented east–west and north–south.

The time required to obtain a radio image of the Sun is determined by the time for the Sun to pass through the diffractional maximum of the antenna beam, or 2.5–3.0 minutes. The possibility of obtaining hundreds (up to 300 in Summer) of radio images over the course of a day makes it possible to trace spatial and temporal properties of the development of events and processes occurring in active regions.

The antenna feed and receiver system distinguishes the total intensity and circularly polarised radio radiation of the Sun. Therefore, two-dimensional distributions of these two Stokes parameters are measured simultaneously.

Part of the antenna system (oriented East–West) is shown in Fig. 11.2, in 1981–1982, when it was being constructed.

The diameter of the surface of each antenna was 2.5 m; two schemes for mounting the antennas were foreseen: a symmetrical cross-like arrangement with resolution $20 \times 2''$, and a T-like arrangement with resolution $10 \times 20''$.

A method for processing radio images using *a priori* information was developed, based on the relationship between the structure of local sources of radio emission in the corona and groups of spots visible in unpolarised and polarised light. This processing makes it possible to appreciably increase the effective resolution of the SSRT.

An automatic tracking system enables the antennas to follow the Sun continuously from East to West at elevations of more than $10°$ above the horizon.

Test recordings of the radio emission of active regions on the Sun using the first group of antennas in the Western arm were begun in March–April 1981. Figure 11.3 shows a series of scans of the Sun obtained from 03^h25^m to 03^h50^m on August 26, 1981 in polarised emission (upper curve) together with a recording of a solar flare in intensity detected at this same wavelength using the radio polarimeter (lower curve).

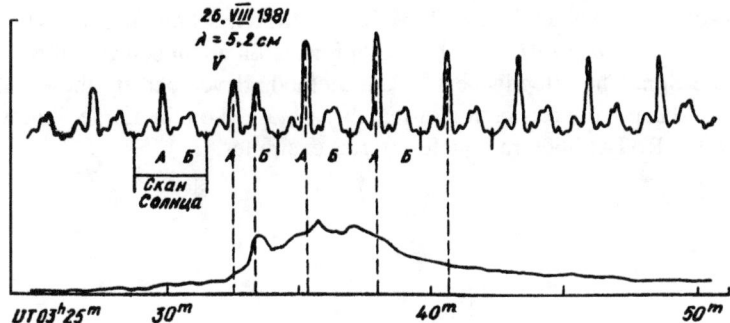

Fig. 11.3 Example of a scan across the solar disk at a wavelength of 5.2 cm (August 26, 1981): (1) using the first group of 16 antennas of the western beam of the SSRT and (2) using a single antenna. The *text below the first pair of letters on the left* "A B" reads "Scan of the Sun"

A solar radio eclipse was successfully observed at nine wavelengths using individual antennas of the SSRT on July 31, 1981, jointly by radio astronomers from the Main Astronomical Observatory, Special Astrophysical Observatory and the Leningrad Polytechnical Institute.

11.3 Research Directions

Three research directions were established and subsequently became traditional in the SibIZMIRAN Radio Astronomy Department.

(1) The development of new radio-astronomy equipment and radio telescopes, first and foremost the SSRT (Smol'kov, Treskov, Krissinel', Potapov, V. V. Kotovich, Putilov, V. V. Belosh, L. M. Risover, V. G. Zandanov and others).
(2) The measurement of characteristic physical processes and events occurring in the solar atmosphere using radio polarimeters, the short-baseline radio interferometer and multi-antenna radio interferometers (V. P. Nefel'ev, Zandanov, Treskov, Treskova, Potapov, Smol'kov, G. N. Zubkova), in close association with optical observations.
(3) Theoretical interpretation of observed processes and events, the elucidation of mechanisms for their origin and development, the construction of physical models and diagnostics for fluxes of energetic particles (A. V. Stepanov, A. M. Uralov, V. G. Ledenev, B. I. Lubyshev, V. M. Bardakov).

The results of solar-physics research carried out by Irkutsk radio astronomers have now become well known both in our own country and abroad. They are systematically published in publications of SibIZMIRAN and in astronomy journals. In the 1983 monograph put out by the *Nauka* publishing house *Methods for the Construction and Processing of Solar Radio Images*, Irkutsk radio astronomers (S. D. Kremenetskii, Putilov, Risover and Smol'kov) develop in detail a variational theory for deriving the fine structure of solar radio images using radio telescopes with high spatial resolution, such as the SSRT of SibIZMIRAN. Much attention was paid to the specification of *a priori* astrophysical information about sources of radio emission on the Sun. The effectiveness of the methods developed by these authors is demonstrated using the processing of real data obtained on the Large Pulkovo Telescope and the RATAN-600 radio telescope as examples.

Chapter 12
Radio Astronomy at the Radio Astrophysical Observatory of the Latvian SSR Academy of Sciences

A.E. Balklavs

Abstract The history of the development of radio astronomy in Latvia is briefly described. Studies were directed primarily at solar activity, including quasi-periodic radio-brightness fluctuations. The results of theoretical studies of mechanisms of solar activity are also noted.

Radio astronomy research in the Academy of Sciences of the Latvian SSR began in 1958 at the initiative of the founder and first director of the Radio Astrophysical Observatory, Dr. J. J. Ikaunieks (Fig. 12.1). The first instrument to be constructed was a radio telescope with a cophased antenna for observations of solar radio emission at 210 MHz (Fig. 12.2).

This radio telescope was included in the programme for the International Geophysical Year (July 1957–December 1958) and the Year of International Geophysical Collaboration (1959). Studies of the flare activity of the Sun that were important for elucidating behaviour of Sun–Earth connections were begun in these years.

During the International Year of the Quiet Sun (1964–1965), radio astronomers of the Radio Astrophysical Observatory were entrusted with the collection and preliminary calibration of all data from radio observations of the Sun in the Soviet Union (N. P. Tsimakhovich). The material collected during the International Geophysical Year and International Year of the Quiet Sun formed the basis for the development of methods for forecasting the near-Earth effects of fluxes of solar plasma from radio observations of the Sun in four frequency ranges. In 1970, it was decided to concentrate efforts and resources on elucidating the behaviour of particle-producing solar activity and on developing techniques for forecasting this activity, based on solar radio observations.

With this goal in mind, the RT-10 radio telescope, with a 10-m parabolic dish and receivers at frequencies of 326, 612 and 755 MHz, was designed and constructed at the Radio Astrophysical Observatory at the beginning of the 1970s (A. R. Avotinykh, A. E. Baklavs, G. A. Ozolin'sh, M. K. Eliass, Tsimakhovich; Fig. 12.3). Starting in 1973, regular observations of quasi-periodic fluctuations of the solar radio emission have been made in close collaboration with the Radio Physical Re-

Fig. 12.1 Founder and first director of the Radio Astrophysical Observatory of the Academy of Sciences of the Latvian SSR Dr. Jan Janovich Ikauineks (1912–1969)

Fig. 12.2 Radio telescope with a co-phased antenna (210 MHz)

search Institute in Gorkii (NIRFI). Observations at centimetre wavelengths (3 cm, with the participation of NIRFI) revealed variations in the character of the intensity fluctuations and the slope of the solar radio spectrum before proton outbursts. Similar phenomena were also seen at a frequency of 350 KHz in observations made from a spacecraft. The relationship between the characteristics of fluctuations in the solar radio emission at decimetre wavelengths and solar activity remained unclear until the Radio Astrophysical Observatory observations.

Research carried out at the Radio Astrophysical Observatory enabled the reliable detection of quasi-periodic components varying with periods from 1–23 minutes in

Fig. 12.3 10-m radio telescope with a fully steerable parabolic surface (operating at 326, 612 and 755 MHz)

the solar radio emission at frequencies of 755 and 612 MHz (E. A. Aver'yanikhina, Ozolin'sh, M. E. Paupere, Eliass and others). Relationships were established between various characteristics of fluctuations of the solar radio emission and radio outbursts associated with solar flares: an increase in the amplitude of these fluctuations by 1–2.5% several hours before an outburst, spectral variations, the appearance of low-frequency pulsations. This was especially clearly seen in proton outbursts observed during August 18–26, 1979, when two regions of proton ejection associated with large-scale coronal magnetic fields were present simultaneously on the solar disc.

A number of characteristic variations in the quasi-periodic fluctuations after the outburst were also discovered (reconstruction of the spectrum of oscillations with a period of 5 minutes, change of the sign of the asymmetry of the fluctuation probability-density function from negative to positive, the appearance of simultaneous ejections at 755 and 612 MHz, corresponding to high fluctuation amplitudes). An operative processing of the observational results enabled forecasting of outbursts 1.5–2 hours beforehand.

The statistical material from observations of radio fluctuations at decimetre wavelengths collected to date remains unable to unambiguously answer a number of questions: Are the effects indicated above associated only with proton outbursts? Are the observed low-frequency pulsations quasi-periodic? What determines the duration of these pulsations? Elucidation of these questions is one of the main research goals of radio astronomers at the Radio Astrophysical Observatory.

In recent years, a number of theoretical research directions concerning various aspects of manifestations of solar activity have been successfully developed at the Radio Astrophysical Observatory: wave motions of the solar plasma, which are closely related to quasi-periodic fluctuations of the solar radio emission (V. A. Lotsans), and modelling the dynamics of the plasma in elementary flare events, which represent an important part of the flare process (A. R. Spektor). The transport of energy by Alfven waves propagating through the solar atmosphere due to the tunnel effect has been studied, and general expressions for the phase velocities of propagating perturbations obtained. It was shown that the phase velocities for the propagation of the plasma motions and of perturbations of the magnetic field differ. The frequency-dependent transmission coefficient for the propagation of Alfven waves through a boundary with a density jump was calculated. The propagation of Alfven waves from the convective zone into the corona was also considered. The dependence of the transmission coefficient for the propagation of Alfven waves through a coronal arch on the frequency and the amplitude of the waves in the arch was also calculated. It was demonstrated that coronal arches act for Alfven waves like high-quality interference filters, and that the interference of Alfven waves in arches can lead to additional heating. The characteristics of the chromospheric plasma in elementary flare events subject to incoming fluxes of accelerated particles and heat was also calculated. It was shown that, when accelerated electrons impinge on the chromosphere, the upper part of the chromosphere can be headed to temperatures of 10^7 K, giving rise to a powerful, short-duration ultraviolet flare in deeper layers, which absorb this energy and rapidly re-radiate it. A narrow flare transition layer forms in the chromosphere. Immediately below this layer, the action of gas-dynamical motions and thermal instability leads to the condensation of cool material, which descends in the chromosphere with an initial speed exceeding the sound speed. This, in turn, leads to the formation of a shock wave that moves in denser chromospheric layers and eventually decays.

The Radio Astrophysical Observatory maintains collaborations with a number of leading scientific organisations in our country (FIAN, IZMIRAN, GAO and SAO, NIRFI, IPG and others). Observational data obtained on the RATAN-600 radio telescope at centimetre wavelengths was processed at the Radio Astrophysical Observatory. Variations in the sign of the circular polarisation of the radio emission were analysed with the aim of elucidating the properties of magnetic fields in active regions. This yielded for the first time data on sign changes of the circular polarisation in a broad range of frequencies for a specific feature in a local source of solar radio emission, making it possible to calculate the configuration of the corresponding magnetic fields in the corona (B. I. Ryabov).

Interesting results were obtained from studies of the solar wind using the 22-m radio telescope in Pushchino (FIAN) to obtain observations of scintillations of distant radio sources on inhomogeneities in the interplanetary plasma (G. Blums and others). From 1958 to 1981, radio astronomers at the Radio Astrophysical Observatory published more than 120 scientific papers. The highly qualified group of radio astronomers of the Radio Astrophysical Observatory of the Academy of Sciences of the Latvian SSR productively work on a number of topical problems connected with solar activity.

In conclusion, I would like to make note of the role of Jan Janovich Ikaunieks in organising the Radio Astrophysical Observatory, forming the scientific group, choosing research directions and outfitting the observatory with optical and radio telescopes.

Being himself a stellar astronomer, Jan Janovich became an enthusiastic radio astronomer, and strove to facilitate joint optical and radio studies of Galactic objects.

Having overcome enormous organisational and technical difficulties in the development and construction of modern antennas and receiver technology, Jan Janovich energetically and flexibly used all possible means available to realise his plans. Solar radio astronomy was probably not the only direction he foresaw on this path.

The premature death of Jan Janovich made it impossible for him to fully realise his aspirations.

Chapter 13
Radio Astronomy Research at Leningrad State University

A.P. Molchanov

Abstract The main directions of radio-astronomical studies at Leningrad State University are described. These studies began in 1947, and have included research on solar active regions, as well as methods for measuring the absolute radio flux of the Sun and the statistical properties of the solar radio emission. Studies of solar–terrestrial connections using radio-astronomy methods are considered in some detail.

Radio astronomy studies at Leningrad State University (LGU) were begun in 1947 in the Radio Physics Department, which was headed by L. L. Myasnikov (1905–1972). In 1949, the first observations of the radio emission of the Sun at 3.2 cm wavelength were carried out by a Ph.D. student in this department, A. P. Molchanov (now a Professor and Doctor of Physical and Mathematical Sciences). A radio telescope built by the Department was used to successfully conduct observations during the total solar eclipses of 1952 (in Pitnyak) and 1954 (in Lagodekhi).

These eclipse observations became the beginning of a long series of measurements of the radio diameter of the Sun, determined by the characteristics of active regions and the propagation of the radio emission through the solar disc. Another problem that was studied in the Department was ways to use the radio emission of the Sun for navigation and antenna measurements.

Starting in 1956, research in these areas was continued by Molchanov in the Department of Radio Astronomy of the Main Astronomical Observatory (GAO) of the USSR Academy of Sciences in Pulkovo.

13.1 Astrophysical Research

In 1960, the radio astronomy studies of V. G. Nagnibeda and R. I. Enikeev were begun in the Department of Astrophysics (headed by Corresponding Member of the Academy of Sciences V. V. Sobolev) of the Faculty of Mathematics and Mechanics of LGU and the Astronomical Observatory of LGU; this work was supervised by Molchanov from 1960 through 1966. They were further supervised by G. P. Apushinskii (until 1971), Nagnibeda (until 1977) and G. B. Gel'freikh (to the present). In 1963, radio astronomy first became available as a subject of specialisation, which has continued to graduate several specialists in this area each year to the present day.

The main research area in the growing group was the radio emission of active regions on the Sun (Molchanov, A. S. Grebinskii and L. V. Yasnov), the development of methods for absolute measurements of the solar radio flux (Nagnibeda) and the statistical properties of the solar radio flux (A. R. Abbasov).

Researchers in the group (which by that time included Apushkinskii) also took part in observations of solar eclipses: in 1962 (Mali), 1963 (Simushir), 1965 (Manya), 1972 (Chukotka) and 1973 (Cuba). In 1966, 1968 and 1975, observations during eclipses were conducted on the 22-m radio telescope of the Lebedev Physical Institute (FIAN) in Pushchino, on the Oka River.

From 1965 through 1972, Nagnibeda carried out a large number of observations of solar active regions at a wavelength of 2 cm using the Large Pulkovo Radio Telescope, in order to study the spectra of local sources of radio emission at short wavelengths.

Another direction in solar research was pursued by A. N. Tsyganov and N. A. Topchilo under the supervision of Apushkinskii: studies of the millimetre-wavelength radio emission of filaments and protuberances.

Today, researchers of the Astronomical Observatory of Leningrad State University participate in observations of solar active regions on the RATAN-600 radio telescope (see Essay 5).

In recent years, studies of the structures of radio galaxies and quasars at metre wavelengths have begun in the Astronomical Observatory.

13.2 Studies of Solar–Terrestrial Connections

In 1966, radio-astronomy research again began to be carried out in the Radio Physics Department in the Faculty of Physics of LGU (headed by Prof. G. I. Makarov). The Laboratory of Cosmic Radio Emission, supervised by Molchanov, was formed, where studies of the radio emission of the Sun were conducted by Abbasov, Grebinskii, Yasnov, I. E. Pogodin and S. A. Andrianov.

Observations by researchers in the Laboratory (V. V. Voronov, L. G. Ipatova, O. V. Korobchuk, S. V. Makaronov, A. P. Sedov, A. G. Stupishin) are carried out on various radio telescopes constructed by V. A. Stupin and Yu. V. Nikitin, as well as on a specialised radio telescope designed for observations on-board a ship. In addition, observations on various large radio telescopes operated by other organisations are carried out as part of collaborative programmes: the 64-m telescope at Medvezhyikh Ozerakh (Bear Lakes) and the 22-m RT-22 radio telescope of FIAN. Collaborative work is also carried out on the processing of observational data obtained on the RATAN-600 radio telescope and the Large Pulkovo Radio Telescope.

Studies of active regions and the distribution of the radio intensity over the solar disc were conducted by Molchanov, Grebinskii, Yasnov, Stupishin and Voronov using observations obtained during solar eclipses in 1966 (Tuapse), 1968 (Kurgan), 1970 (Cuba, Mexico), 1973 (Mauritania), 1975 (Batabad, Riga), and 1981 (Dushanbe, Kamen'-na-Obi), as well as on large radio telescopes, as part of collaborative programmes, in 1975 (the 22-m telescope of FIAN, the 64-m Medvezhikh Ozerakh telescope).

A large series of investigations of the spectral characteristics of the radio emission of various regions on the Sun was carried out on the 22-m FIAN telescope, jointly by researchers in the Laboratory and FIAN. This same telescope was used during observations of the transit of Mercury across the solar disc on May 9, 1970. Yasnov, Korobchuk and other studied the fine structure of the radio emission of active regions, and later, fluctuations of the radio emission of solar active regions (Molchanov, Yasnov and others).

Systematic observations of solar eclipses (including observations conducted at the Astronomical Observatory and the Main Astronomical Observatory in Pulkovo) enabled Molchanov to determine the radio diameter of the Sun and its dependence on wavelength (at centimetre and decimetre wavelengths) and the phase of the solar activity cycle. He was also able to show that the thickness of the transition region in the solar atmosphere is modest (Molchanov, Yasnov). The distribution of the radio intensity across the solar disc was also studied (Molchanov, Greginskii, Yasnov and others).

Research in the radio emission of active regions on the Sun was pursued in three main directions.

(1) A statistical study of the characteristics of active regions enabled Abbasov, Grebinskii and Korobchuk to investigate their relationship to optical parameters and to identify quasi-periodic components in the solar radio emission (with periods of more than 10 days). Methods for the statistical forecasting of slowly varying components were also developed (Andrianov, Yasnov and others), as well as the transition-matrix and evolutionary-curve methods (Grebinskii, Korobchuk), which were successfully used for day-to-day forecasting. These forecasting methods represented a substantial improvement over methods used earlier in practical applications of solar radio measurements.

The statistical studies also made it possible to try to forecast solar flares and their consequences in near-Earth space (Molchanov, V. Yu. Osadchii). Pogodin compiled a review of forecasting methods, as well as a catalogue of proton events in 1970–1977. The conditions for the propagation of cosmic rays in interplanetary space were investigated, and a method to take into account these conditions when forecasting the consequences of flares was developed (Pogodin and others). The practical realisation of the results of these studies was the creation of so-called single-wave forecasting of the consequences of solar flares in near-Earth space, which yielded satisfactory results and is now being verified by researchers of the Laboratory with the participation of colleagues at the Institute of Applied Geophysics.

(2) The second main research direction is studies of physical processes determining the radio emission of active regions. In this area, the results of statistical studies enabled Pogodin to elucidate a whole series of interesting properties of outbursts, and to develop a model for the development of flares in the solar atmosphere.

Independent of this, Grebinskii and Sedov developed a flare model taking into account the Razin effect, proposed a method for determining the magnetic fields in the region where the flare originates and considered the possible role of non-thermal mechanisms in the generation of radio emission by local sources on the Sun. These researchers also conducted theoretical studies of the relationship between the X-ray and radio emission of the Sun. Yasnov, Ipatova and others considered the scattering of radio waves on plasma-turbulence waves, the propagation of high-frequency plasma waves perpendicular to the magnetic field taking into account relativistic effects, the generation of high-frequency plasma oscillations by beams of particles, the propagation and stability of magnetohydrodynamical waves, and the interaction of magnetoacoustic waves and active regions on the Sun, as well as a quasi-linear theory of the growth of Alfven waves in an inhomogeneous medium and the possible formation of radio spectral lines in active regions.

The information obtained about these various physical processes made it possible to refine models of active regions and to make real use of these data to improve the single-wave method for forecasting consequences of solar flares.

(3) The third research direction is introducing the results of various studies and developed methods into practice: regular measurements of the solar radio flux and of the effective location of its centre are carried out, and forecasts made; to realise the proposed method for forecasting the consequences of solar flares, a method for observing the solar radio emission using a telescope with a moving base without the need of a gyroscope was developed and put into practice (in collaboration with the Institute of Applied Geophysics).

Methodological work is also carried out in the Laboratory. Methods for determining the shift in the effective centre of the radio Sun were refined, new aperture-synthesis methods and methods for the determination of the coordinates of flares were developed, and questions concerning the calibration of data during solar radio measurements were investigated. In addition, The Laboratory researchers Andiranov and Korobchuk actively took part in the development of mathematical tools associated with observations on the RATAN-600 radio telescope.

Index

1–9
104α, vii
21-cm radio line, 90
22-m radio telescope, xvii
3C111, 150
3C120, 165
3C147, 43
3C196, 43, 197
3C273, 57, 96, 103, 165, 166
3C48, 43, 165
3C84, 165
6-m optical telescope, 109
70-m radio telescope, xix, 39, 59, 78, 98, 107, 108, 214

A
Absorption, 200
Acousto-optical spectrometer, 147
Active galaxies, 53
Alfven waves, 239
Alupka, 15
Alushta, 15
Ambartsumian, v, xi, 99, 157
Angular structure, 192, 194
Annihilation line, 54
Aperture synthesis, 76, 213
Aperture-frequency synthesis, 85
Arc method, 153
Artificial moon, viii, 73
Artyukh, 43
Astrometric research, 154
Astronomical unit, 216

B
BAO, 157, 161
Barret, 40
Bistatic radar, 222
BL Lac, 48
BL Lac objects, 106
"Black" disks, 72, 74
Borodzich, vii, 40
Bracewell, xii
Braude, vi, ix, 181
Bremsstrahlung, 195
Brewster angle, 73
Brightness temperatures, 106
Broadband Cross Radio Telescope, 36
BSA, 46–48
Byurakan Astronomical Observatory, 43, 48
Byurakan Astrophysical Observatory, 99, 157

C
California Institute of Technology,, 39
Callisto, 42, 149
Carbon, 200
Cas A, xi
Cassiopeia A, 68, 84, 95, 158, 163, 165, 182, 197
Cassiopeia A, Cygnus A, Taurus A, Virgo A, 79
Centaurus A, 80, 150, 164
CH, xi, 23
CH line, 91
Chesalin, 42
Chikhachev, vi, 3, 13
Chromosphere, 89
Circular polarisation, xvii, 177
Clark, 39
"Clean" map, 147
Closure phases, 154
CN, 103
CO, 83
Comet Kohoutek, 50
Commission for Radio Astronomy, xix

Compton losses, 54, 202
Confusion, 185
Convection, 55
Corona, 89, 229
Correlation radiometer, 96
Cosmic masers, 84
Cosmic microwave background, xi, 82, 128
Cosmic radio sources, 38, 178, 192
Cosmic rays, 52, 54, 55, 91
Cosmogony, 93
Cosmos-669, 58
Crab Nebula, xvii, 10, 18, 20, 39, 40, 43, 68, 71, 92, 94, 103, 196
Crab supernova (Taurus A), 80
Crimean Astrophysical, 42
Crimean Astrophysical Observatory, 83
Crimean Station, 15
Cross radio telescope, 43
CTA 102, vii, xi, 96
CTA 21, xi
Cyanacetylene, 51
Cyanamide, 83
Cygnus A, 96, 150, 163, 164, 182
Cygnus X, 24

D
Dagkesamanskii, 43, 44
Decametre wavelengths, 184
Deep Space Communications Centre, xix, 26, 40
Deuterium, xi, 90, 122, 131
Dielectric permeability, 211
Diffractive scintillation, 57
Diffusion, 55
"Dirty" map, 147
Discrete radio sources, 194
Discrete sources, xviii, 79, 130
Dispersion interferometry, 222
Dispersion measure, 107
DKR-1000, xvii, 34, 42, 45, 51, 108, 165, 168
Doppler beaming, 109
Doppler effects, 97
Doppler shift, 200
Doroshkevich, 103, 128
Double radio sources, 97
Dravskikh, vii, 131, 153
Dynamical spectra, 79

E
Eclipse, 43
Effelsberg, 108
Electron bolometer, 103
Electron density, 199
Electron temperature, 199

Energy losses, 96
Ephemerides, 207, 210, 214
Europa, 149
Excited hydrogen, 40, 42
Excited-hydrogen line, 131
Extragalactic sources, 35
Extraterrestrial civilisations, 98
Extraterrestrial intelligence, viii

F
Faraday, 68
Faraday rotation, 21, 22, 47
FIAN, vi, 29
Formaldehyde, xviii, 147
Fourier interferometers, 59
Fraunhofer, 85
Fraunhofer zone, 79
Fresnel, 85
Fresnel zone, 79

G
Gabuzda, xii
GAISH, x, 89, 93
Galactic Centre, 53
Galactic spur, 81
Galaxies, 194
Gamma-ray radiation, 53
Ganymede, 149
Geomagnetic phenomena, 48
GIFTI, 61
Ginzburg, v, x, 1, 3, 51, 89
Global VLBI, 107
Goldstone, 106
Gorelik, 8
Gorkii, vi, viii, 8, 13, 65
Gorkii Physical–Technical Institute, 13
Gorkii State University, 61
Gorozhankin, 9
Gravitational energy, 53
Green Bank, xii, 106
Gyrofrequency, 128

H
H 56α radio line, 42
H_2CO, 152
H_2O, 106, 152
H 104 line, 41
H56α, xviii
H90α, vii
Halo, 54
HCN, 84
HI, xi, 151
HI absorption, 200
High-frequency excesses, 41
H(II), 195

Index

H II regions, 91, 194, 199
Hubble constant, 111
Hydrogen, xviii, 122, 147, 171
Hydrogen line, 23
Hydroxyl, xviii, 147
Hydroxyl line, 90

I
IKI, 42, 89
Infrared, 53
Inhomogeneities, 55
Interferometry, 84
Interkosmos, 220
International Geophysical Year, 23, 24, 231, 237
International Year of the Quiet Sun, 24, 237
Interplanetary plasma, 46, 48, 168
Interstellar medium, 47, 107
Interstellar molecules, 90
Interstellar plasma, 45
Interstellar water-vapour, 161
Io, 149, 198
Ionised gas, 130
Ionised hydrogen, 184, 199
Ionosphere, 86, 185, 192
IRFE, 181

J
Janovich Ikauineks, 238
Jets, 53
Jodrell Bank, 46
Jodrell Bank Observatory, 213
Jupiter, 79, 83, 130, 198, 202, 212

K
Kalachev, 11, 13, 26, 50
Karadag Radio Astronomy Station, 73, 75
Kardashev, v, vii, 14, 24, 92, 95, 101, 107
Katsiveli, 176
Keldysh, 38, 44
Kepler supernova remnants, 80
Khaikin, vi, x, xix, 4, 13, 15, 26, 61, 119, 226
Kharkov, vi, ix, 79
Kislyakov, 82
Knife beams, 126
Kogan, 42
Konovalenko, ix
Korol'kov, 154
Koshka, 9
Koshka Mountain, 15
Kostenko, 42
Kotel'nikov, vi, viii, xv, xix, 38, 205
Kraus-type radio telescope, 99
KRT-10, 105
Kuz'min, vi, xii, 22, 28, 38

L
Large Pulkovo Radio Telescope, xvii, 121, 137, 183
Large Pulkovo Telescope, 236
Lebedev Physical Institute, x, 15, 183
Leningrad State University, 243
LGU, 243
Limb-darkening, 218
Linear polarisation, xvii, 21, 22, 47, 81
Litvinenko, ix
$\log N - \log S$, 151
Long-baseline interferometers, 47
Longair, xii, 111
Lovell, 213
Lozinskaya, 24
LPRT, 121, 137
Lunar occultation, 80
Lunar radio emission, 27, 70
Lunar "seas", 38

M
M-6000 computer, 49
M87, xi, 95
Magnetic bremsstrahlung, xvii
Mandel'shtam, xv, 1
Mariner-2, 130
Markarian, 166
Markarian galaxies, 48, 166
Mars, 83, 212, 214
Maser emission, 90
Matveenko, vi, 23, 26, 40–42, 106
Mercury, 83, 152, 212
Metagalaxy, 184
Meudon Observatory, 102
Micropulses, 47
Moiseev, 42, 172
Molecular radio lines, 50
Moon, 32, 71, 78, 130, 205

N
Nancay telescope, 102
Neutral hydrogen, 23, 162
Neutron Stars, 56
NGC 3034, 103
NGC 7027, 103
Nitrogen, 90, 200
Nodding, 66
Non-thermal, 52
Novikov, v, x, xi, 103, 128
NRAO, 83

O
Occultation, 22, 71
OH, xi, 102, 152

OH maser, 106
Omega Nebula, 40, 131
Ootakamoond, 160
Orion, 131
Orion A, 152
Orion arm, 200
Orion Nebula, 116
Oxygen, 69
Ozernoi, 53

P
Pancakes, 111
Papaleksi, v, xv, 1
Paramagnetic amplifiers, 178
Parametric amplifier, 127, 147, 162
Pariiskii, v, vi, 27, 92, 134, 137
Penzias and Wilson, xi
Perseus A, 164
Perseus arm, 200
Phased array, 45
Pikel'ner, v, 90, 94
Planets, 194
Plasma, 56, 226
Polarimeter, 124, 228
Polarisation, 19, 22, 45, 68
Pravda, vii
Proton outbursts, 239
Pulkovo, vi, x, 165
Pulkovo Observatory, 118
Pulsar, 197
Pulsar profile, 47
Pulsars, xix, 44, 46, 56, 108, 194
Pushchino, vi, vii, x, 15, 26, 28, 30, 40, 42, 96, 174, 175

Q
Quasars, 48, 53, 95, 106, 109, 165, 166, 194

R
Radar, 205
Radar astronomy, xviii
Radar mapping, 213
Radial velocities, 214
Radio astrometry, 85
Radio Astronomy Laboratory of FIAN, 30
Radio Astronomy Station of FIAN, 30
Radio galaxies, 95, 106, 166
Radio granulation, 148
Radio heliograph, 233
Radio outbursts, 239
Radio Physical Research Institute, 71
Radio recombination, vi, 95
Radio recombination lines, vii, xviii
Radio spectra, 41

Radio spectrograph, 173
Radio stars, 52
Radio-Physical Research Institute, vi
Radio-Relay-Linked Interferometer, 42
RadioAstron, ix
Radiospectrographs, 78
RATAN-600, x, xii, xviii, 80, 100, 132, 134, 137, 167, 236, 244
Razin, 64, 68
Recombination lines, ix, xviii, 199, 200
Refraction, 192
Relativistic electrons, 52
Relict radiation, 103, 109
Remote sounding, 220, 221
Rotation measures, 80
RT-22, 28, 37, 51
RT-70, 107
RT-MGU, 99

S
Sagittarius A, 165, 182
Sagittarius B2, 152
Salomonovich, 9, 11, 26, 38
Salyut-6, 105
Sanamian, 157
SAO, 134, 137, 153
Sazonov, 53
Scattering, 54
Scientific Council for Radio Astronomy, xv
Scientific Council of the USSR Academy of Sciences on Radio Astronomy, xix
Scintillation, 43, 44
Scintillation index, 168
Sea interferometer, 16
Sea radio interferometer, 10
Search for extraterrestrial civilisations, xviii
Seismic, 85
Self-absorption, 195
SETI, xi
Severnii, 171
Shain, 9
Shavkovskii, 38
Shklovskii, v, 3, 23, 89, 109, 158, 200
Shmaonov, xi, 128
Shock fronts, 55
Sholomitskii, vi, vii, 96
Simeiz, 15, 106, 174
Slysh, v, x, 98, 101
Soboleva, 92
Solar atmosphere, 57
Solar corona, 56, 68, 182
Solar flares, 57
Solar noise storms, 177
Solar plasma, 237

Index

Solar radio emission, 171
Solar radio flares, 177
Solar Service, 72
Solar wind, 44, 48
Sorochenko, vi, vii, 10, 18, 23, 24, 40, 42, 131
Space interferometers, 104
Space radio telescope, 98
Space Research Institute (IKI), x, 42, 47, 89, 101
Special Astrophysical Observatory, 134, 137
Spectral index, 195
Spectrograph, 228
Sramek, 167
Standard sources, 67
Star formation, 107
Stark, vii
Stark broadening, vii
Sternberg Astronomical Institute, v, vi, 89
Submillimetre, 58
Sun, xvii, 61, 171, 194
Sunyaev, v, x, 110–112, 116
Sunyaev–Zel'dovich effect, 111
Supercorona, 20, 31, 34, 43, 44
Supergranules, 177
Supernova remnants, 46, 80, 91, 94, 95, 199
Supernovae, 194
Swenson, xii
Synchrotron radiation, 21, 52, 91
Synchrotron reabsorption, 98
Synchrotron self absorption, xi
Syrovatskii, 51

T

Tata Institute, 169
Taurus A, 10
Tectonic, 85
Tectonic activity, 224
Thermal radio, 42
Thermal radio emission, 220
Time variations, 96
Tovmasian, 24, 167
TPA/I mini-computer, 50
Travelling wave tube, 70
Troitskii, vi, viii, 13, 62, 69, 73
Tropheric turbulence, 154

Tropospheric absorption, 85
Tula Polytechnic Institute, 49
Tycho Brahe, 80
Type I, Type II, and Type II civilisations, xii

U

Udal'tsov, 22
Ukrainian SSR, 181
URAN, 193
Uranus, 42
UTR-1, 188
UTR-2, 79, 185, 190
UTR-2 radio telescope, xviii

V

Van de Hulst, vii, 90
Variability, 53
Variable-profile antenna, 119
Venera, 223
Venus, viii, xii, xvii, 38, 39, 42, 83, 130, 205
Very Long Baseline Interferometry (VLBI), ix, xviii, 42, 85, 137, 176, 193
Virgo A, 182
Vitkevich, vi, xvii, 9, 13, 15, 21, 23, 25, 43, 44, 168, 226
VLBI, 42, 44, 106, 137, 153, 159, 176, 179
VPA, 119, 148

W

W49, 103
W51, 103
Water maser sources, 48
Water vapour, xviii, 69, 85, 147
Westerbork Synthesis Radio Telescope, 167

X

X-ray astronomy, 109

Y

Year of the Solar Maximum, 229

Z

Zel'dovich, v, x, xi, 110
Zimenki, 72
Zimenki Station, 63

GPSR Compliance
The European Union's (EU) General Product Safety Regulation (GPSR) is a set of rules that requires consumer products to be safe and our obligations to ensure this.

If you have any concerns about our products, you can contact us on

ProductSafety@springernature.com

In case Publisher is established outside the EU, the EU authorized representative is:

Springer Nature Customer Service Center GmbH
Europaplatz 3
69115 Heidelberg, Germany